'A *stirring* rural *fantasia* . . . Lewis-Stempel's heart and mind are absolutely in the right place. I salute him and I adored his appreciation of the *quirky detail*.'

Roger Lewis, *The Times*

'John Lewis-Stempel knows not only all about the different kinds of life in such a place and how they all fit together, but can also *write so vividly*.'

Philip Pullman, *Guardian*

'John Lewis-Stempel describes beautifully the changing of the seasons and the habits of animals such as the hares that make their home in his field. The book is *a superb piece of nature writing*.'

Ian Critchley, *Sunday Times*

'There *is a raw honesty to this book* . . . a sense of reviving what was thought to be lost, that sustainability may be buried deep but is nevertheless still there; like King Arthur, Old England is not dead but is slumbering and can awake and return.'

Nick Groom, *The Times Literary Supplement*

'*The Running Hare* is *funny, erudite and a delight* from start to finish. John Lewis-Stempel knows the land, loves it — and works it. He is a farmer, muddy-booted and diligent, who effortlessly recreates on the page the intimacy with the natural world that his daily rounds bring. But farming is also the enemy in his piece — the farming of the mega-tractor and the sprayer, the farming that has, during his lifetime, quietly destroyed the greater part of the country's flora and birdlife. *The Running Hare* is *an important book*, as richly layered and rewarding as the soil of an unimproved field.'

Philip Marsden

'*A cautionary tale* of the wasteland awaiting us if nothing is done to halt the chemical warfare being waged on the wildlife that once flourished on farmland.'

BBC Countryfile Magazine

'Fans of Lewis-Stempel's bestselling *Meadowland* will find here the same easy-reading prose fuelled by daft-as-a-brush enthusiasm and embellished with lyrical flourishes . . . *the mud-spattered details of a farming life lend The Running Hare a unique realness.*'

Richard Benson, *Mail on Sunday*

'*I love the earthiness* of John Lewis-Stempel's writing . . . he has done it, done it, sweated it, and has earned the right to write about it with straw-cut fingers . . . Brilliant stuff.'

Christopher Somerville, author of *The January Man*

'Lewis-Stempel's writing is *quietly passionate, intensely descriptive and beautifully detailed*. This book is deeply moving and ultimately, hopeful . . . One of the ten best conservation and environment books of 2016.'

Forbes

'A beautifully observed book, full of poetic descriptions. *Brilliant and galvanising.*'

Charlotte Heathcote, *Sunday Express*

The Running Hare

The Secret Life of Farmland

John Lewis-Stempel

BLACK SWAN

TRANSWORLD PUBLISHERS
61–63 Uxbridge Road, London W5 5SA
www.penguin.co.uk

Transworld is part of the Penguin Random House group of companies
whose addresses can be found at global.penguinrandomhouse.com

Penguin
Random House
UK

First published in Great Britain in 2016 by Doubleday
an imprint of Transworld Publishers
Black Swan edition published 2017

Illustrations by Micaela Alcaino

Lyrics from 'Combine Harvester' (O'Shaughnessy/Safka) reproduced by
kind permission of Sony/ATV Music Publishing, EMI Music Publishing.
All rights reserved.

Every effort has been made to obtain the necessary permissions with
reference to copyright material, both illustrative and quoted. We apologize
for any omissions in this respect and will be pleased to make the
appropriate acknowledgements in any future edition.

This book is substantially a non-fictional account based on the life, experiences and
recollections of the author. In some cases names of people or the detail of events
have been changed solely to protect the privacy of others. The author has stated to
the publishers that, except in such minor respects not affecting the substantial
accuracy of the work, the contents of this book are true.

A CIP catalogue record for this book
is available from the British Library.

ISBN 9781784160746

Typeset in Goudy by Falcon Oast Graphic Art Ltd.
Printed and bound by Clays Ltd, Bungay, Suffolk.

Penguin Random House is committed to a sustainable
future for our business, our readers and our planet. This book
is made from Forest Stewardship Council® certified paper.

MIX
Paper from
responsible sources
FSC® C018179

5 7 9 10 8 6

To the brown hare, the corncrake, the poppy,
and the partridge (grey and red-legged).

Contents

Preface

Now that I look back, I see that I have written with some anger.

It happened like this. A few years ago a friend suggested that we, in a mutually beneficial arrangement, grazed our sheep in her paddock. We got the sheep fed, she got the grass cut.

She lived in lowland east Herefordshire, a place I knew well from childhood, and in tending the sheep there I was forced to leave the wonderland in which I lived, the Black Mountains of the England–Wales border. Up on those dark but heavenly hills skylarks sang, otters swam in the brook, and polecats eyed up the chickens.

Where our friend lived was beautiful, but as life-full as a cemetery. Someone had removed the birds from the farmland all around her. For hundreds and hundreds of square miles around her.

The Farmer is to blame. The Supermarket too. And let us not forget the Politician, and the Consumer. Let us not omit Me, or You.

Really: I just want the birds back.

CHAPTER I

I Take This Field

Soil. Earth. Dirt.
Call it what you will, it's the element
of agriculture, of farmland . . .

I WAS EARLY. A parent being early is as useless as a
parent being late, but it was only when I checked the
texted directions on my phone that I realized it said '2',
not '12', for post-sleepover pick-up. What to do for two
hours while I waited to prove my paternal reliability and
save her teenage embarrassment? I was near British
Camp on the Malvern Hills, those dinosaur-spine eruptions into the cultivated English Eden, and it was years
since I'd last been up, so up I went, blue coat flapping
like a sheet on the line. Edward Elgar lived close by for
a decade, so 'Nimrod' played loud in the air as I ascended
to the Iron Age fortress.

Breathless, at the top I sat down for the view, which
in another way took my breath away. Laid out, as in the
view from an aeroplane window, was Herefordshire,

the whole of it, to the Black Mountains in the west, the shining Wye to the south, the Clee Hills to the north.

This is my heartland. Once, my London-born wife asked me to mark on a map everywhere my family, both paternal and maternal lines, have been born. From here I can see every place for the last eight hundred years. She laughed, but kindly, with the appreciation of someone whose own family have wandered.

It was warm in the August sun and I was tired, so I lay down in a hollow and fell to drowsy dreaming:

Dream I
A memory, actually, from some time in the 1970s, I can't be sure when, but before the river of life hit the dividing rock of exams, when some went one way, the rest elsewhere:

I finger-toe climb the gappy stone wall behind my grandparents' house in Herefordshire (going through the gate would be no Everest adventure) into the wheatfield. The cereal is gold and heavy-headed, the evening sun blood-red, the scene a Stalinist painting of promised-land plenty. I start pushing through the rows of the crop; since I am small and the wheat tall (wheat was dwarfed soon afterwards so it did not bend under the weight of chemical sprays) I can hold my arms aeroplane-like and skim the hard heads with flat hands to achieve equilibration. There is a slight wind in the wheat; my hands and the breeze make sibilance.

Above me, and in fancy, swallows are Spitfires wheeling and diving.

I stumble, look down, and put away the childish game. There are poppies and cornflowers and corn marigolds weaved through the cornstalks; and in the bare earth circle, where the seed drill faltered, a cowering grey bird.

I know what it is instantly, because I have spent days poring over bird books, trying to identify the bird making the comb-scraping noise. I've asked my grandparents; 'Rail,' they said, but to me a rail was the moorhen on the farm pond. Finally I had twigged. They meant landrail, or corncrake. The bird with the onomatopoeic Latin name: *Crex crex*.

The corncrake evanesces. Perhaps for a tenth of a second our eyes had met; a lifetime, in other words. Wordsworth once wrote of 'spots of time', experiences so intense they expand and inform existence ever after. They have a 'renovating virtue'.

In that cornfield I looked into the eyes of what was probably the last corncrake in Herefordshire.

I have never forgotten you, corncrake.

I WOKE UP WITH a guilty start from the day-dreaming, thinking I had slept too long, and scrabbled for my phone but found I had only catnapped for minutes. I looked again at the view, at the immense spread of fields, a watercolour paintbox of solid blocks of green and gold. There is a pleasant land before me, but I know when I descend into those fields they are silent, sterile, open-roofed factories for agribusiness. Units of production.

At this point, full disclosure: I farm.

I change the subject in my head to something more agreeable and get in touch with my inner teenage Eng Lit student. Somewhere on these same slopes, William Langland, the fourteenth-century poet, had his character Will fall 'into a slepyng' and meet a spiritual guide, Piers Plowman, who showed him a vision of a just society. I wonder where on the slopes exactly Will slumbered?

I must have dozed off again . . .

Dream II

·Piers Plowman is holding the reins of oxen while declaiming to a group of snaggle-toothed peasants about a good society. I am at the back, taller than the rest, leaning over . . . and he points at me . . . then I see myself back in the cornfield with the corncrake but now I'm middle-aged . . .

I'M WIDE, WIDE awake now. There is no sophisticated, writerly way to put this: I have had a gutsful of chemical farming. If the chemicals dousing the land are so fantastically safe, why do crop-sprayers have sealed cabs? By law, specifically European normative 15695-1:2009, the carbon filtration system on a crop-sprayer must be 99 per cent efficient in preventing any ingress into the cab of toxic dust and vapours. If pesticides and herbicides are dangerous to farmers, they are dangerous. Period.

Now I've got a vision of my own. Piers ploughed in order to ameliorate society's evil. Why don't I take a modern, conventionally farmed arable field, plough it, and husband it in the old-fashioned, chemical-free way, and make it into a traditional wheatfield? Bring back the flowers that have all but disappeared from British ploughlands, such as corncockle, Venus's looking-glass, shepherd's needle, corn marigold and the cornflower, with its bloom as brilliant as June sky? And the birds and animals that loved such land – grey partridges, quail, harvest mice.

And hares. Could I entice in a hare? A corncrake is an impossibility because they are extinct in England except for a small introduced colony in Cambridgeshire. But perhaps I could manage a hare.

Such is the vision of John the Plowman.

There is a problem. We farm in the hills of the far distance under the black wall of Wales, where nothing but grass and sheep grow. I need to find an arable field. A field to plough to grow crops.

Another confession: I grew up with arable farming. And I miss it so.

The moon lies back and reddens.
In the valley a corncrake calls
Monotonously,
With a plaintive, unalterable voice, that deadens
My confident activity;
With a hoarse, insistent request that falls
Unweariedly, unweariedly,
Asking something more of me,
Yet more of me.

D. H. Lawrence, from 'End of Another Home Holiday'

BUT NOBODY WANTS to rent me an arable field to turn into a traditional wheatfield. I advertise, I tweet, I put up cards in village shops from Ross to Ledbury.

There is the problem of taking a field out of crop rotation, there is the bigger problem of the 'W' word. As soon as I mention, to the handful of bites that I do get, that I wish to sow wildflowers in with the wheat I get the same response: 'Those are weeds, they might contaminate our crop.'

One arable farmer of my acquaintance is more succinct still; I'm on my egg delivery round (we have free-range chickens, lots of free-range chickens . . . Light Sussex, Cream Legbars, Araucanas, Marans, Minorcas,

Wyandottes, Speckledys, Barnevelders, Warrens, Old English Game) on the narrow back lane at Wormbridge when I have to slow the Land Rover down to pass an oncoming black Nissan Warrior 4x4, which is only slightly bigger than a battleship. I wind down the window, we chat for a moment, I pop the arable field question, he replies: 'Weeds? You want weeds? I'll show you some f****** weeds . . .'

Eventually, through a friend of ours, Joanna, I'm put in touch with Philip Miller, an advertising executive, who owns some land at St Weonards in south Herefordshire. 'He's a keen birdwatcher,' says Joanna.

A decade ago, on a fancy, Phil Miller bought a three-acre wood; with the woodland came three fields, two permanent pasture, one arable, plus a derelict cottage garden. He lives in St Albans, and rents the fields out. The present tenancy ends in December. Eventually, after some haggling, I take on all the land, fifteen acres in total, on a two-year Farm Business Tenancy. Little of this is ideal, and the least-good aspect is that I can only have wildflowers in the arable field for a year. After that, I have to put the field down to grass.

One year. One opportunity.

THE FIELD HAS a name: Flinders, after an owner of long ago, so I am told. It is four acres in extent, now crammed with kale for livestock to eat in situ. (I have to buy the pert green forage from the last tenant.) Winter comes out in January's false sunshine as I walk around the edge of my new, if temporary, possession, mapping the field in my head.

The field is almost square, crew-cut hedges on three sides, wire stock fence on the other, a bit of ditch underneath this; the ditch then 'dog-legs' to run deep under the western hedge and is swollen thick with red topsoil running off Flinders and the twenty-acre wheatfield belonging to next door. A few spavined nettles hang over the silt snake, a fern or two cling to the ditch side. At first glance, Flinders is a disappointment, an unremarkable field, featureless, an *ager rasus*.

Gnats vortex in the untimely heat. A gang of discordant jackdaws plays some juvenile game in the sky. Otherwise the field is silent; a mausoleum with invisible walls and roof.

I walk around the field again, and peer closer, at everything: the kale, the ditch, the hedges, the two-foot grass margin, muted by winter, which edges the field in a hairy fawn frame. In the west hedge, two trees, alders, have recently been felled, sawdust splattered everywhere. Why? Probably because they were shading the crop. This is a field without a single tree in the hedges. In the top, northwest corner sits a pleasantly stubborn elm stump, sufficiently into the field that the kale is forced to swerve around it.

'Tram tracks', the wheel marks from a crop-sprayer, are ghost lines in the kale. Once upon a time the countryside was criss-crossed by roads, paths and bridleways;

the new tracks across the landscape are guides for machinery.

At second glance, Flinders is a greater disappointment still. As I close the gate behind me, a crop-sprayer trundles past. You may see a fine day; a conventional farmer sees a day to go a-spraying with 'a post-emergence herbicide'.

Only in one respect is Flinders unusual: it is a runt of an arable field. The fields next to it are twenty acres minimum.

I suppose the wide open landscape south and east towards Ross is pleasant enough, as it rises and falls in long swells, and spires of poplar somewhat disrupt the regularized grid of big fields, which replicates endlessly into the eastern distance. In the unseasonal sun alders along a far brook appear as a port-wine stain.

You know how squinting enables you to see a pixelated picture clearly? On some inexplicable impulse I narrow my eyes at the view, and for a passing second I see the faint indentations in the earth where hedges were before the great sixties rip-up. I see the beautiful past.

The farmer-politician William Cobbett travelled this way in 1821 on his *Rural Rides* and averred: 'Everything here is good, arable land, pastures, orchards, coppices, and timber trees, especially the elms, many scores of which approach nearly to a hundred feet in height.'

The elms are long gone; the orchards too. But Cobbett was right; the heavy clay land, given a chance, likes plants and trees to grow upon it. Following the last ice age the earth here gave rise to thick oak forests, and the area's first farmers, the Neolithic people, began their arable farming by wearily cutting down the oaks with their polished axes to make small allotments. Humans have been growing food here for five thousand years. It is small wonder, I think, that the word *human* and the word *humus*, meaning 'soil', come from the same root in Proto-Indo-European, the ancestral language of the Indo-European family. That root is *(dh)ghomon*, meaning 'earthly being'. The Hebrew *adam*, meaning 'man', is from *adamah*, ground.

We live off the earth, and when we die we go back to it. To add to the *humus*.

Within a mile to the west of Flinders the hills begin, and the roads worsen. 'God help us' is the Herefordshire saying about the village of Orcop, partly in relation to the potholed primitiveness of its lanes, partly because it is the first village of the Welsh Borders, that dubious, disputed edgeland.

Flinders is within, by ten clod's throws, the arable Midlands, safe within the pale of civilization. The Romans were here, the Saxons too. Neither of them much fancied the dark wet hills and the Welsh.

Farming is about rain. For arable farming, one needs twenty-five or so inches of rain per annum; thirty-five inches is too much. Up in the mountains where we live, the year's rainfall can easily top fifty sodden inches.

Go west, young man? Hardly. In Herefordshire if

you had made money you progressed eastwards, to the good, dry lands. It is not just a trick of my memory that my 1970s childhood in east Herefordshire was golden; it was actually, physically, meteorologically 50 per cent drier than where I now live, though the distance across the surface of the planet is just twenty miles.

I once contemplated a small hop yard on our hill farm, calculating that I had enough farmyard manure (FYM in the jargon, FYI) to nourish the soil and those sky-seeking hungry bines. 'You'll never sustain it, John,' said Leslie Rees, our neighbouring farmer and knocking on the door of seventy-five. He was right, of course, because he had come from Stretton Sugwas in the east of the county as a farm labourer, and worked (and worked again) to become a farmer and set his own four sons up. You could do that as late as the 1990s, before the City boys started to speculate in agriculture. Gold down? Land up!

Leslie was right. The damp would have blighted the hops, plagued them with mould. I knew it too, because my maternal grandfather was a hop farmer at Much Cowarne. Among my earliest memories is being in a winter hop yard, helping headscarved women pull down the brittle dead bines from towering wires. The bines were put on a ceaseless bonfire; lunch was a potato baked in the ashes.

To have my own hop yard was a dream, just a dream. For years I thought my longing for hops was about the personal sanctuary of childhood, that time before

23

bills and responsibility; and then one toss-turny night I revisited the hop yard of memory, stood in the middle of the place, and swivelled the full 360 degrees. I saw past the people and the wooden pillars to the birds. There were coveys of grey partridge scuttling through.

As I climb into the Land Rover cab, I glance north. The view is blocked by the heavily wooded Aconbury Hill where my father's family, who were Norman come-latelies, started farming in 1450. I cannot escape the shadow of the past, and I do not wish to.

On the drive home, I keep running the word 'Flinders' around my head. Does it not mean something?

Then there is the weather: like others who work land, I am prone to believing folkloric adages. Science works wonderfully in an electric-lit laboratory; the real world is less test-tube certain. It is never a good omen in farming if January begins warm.

January, eh? 'The blackest month in all the year / Is the month of Janiveer.'

ON 4 JANUARY I ship sheep over to Flinders to eat the kale. Backwards and forwards with the Land Rover and trailer, until I've sixty sheep in the field, a 'mob stock' to eat the forage down so I can plough when the weather turns right. Overnight the weather has gone to ice so of course the heater in the cab of the Land Rover breaks down en route. I pull our Jack Russell terrier across my lap as a hot-water bottle.

Sheep: won't go into a trailer, and once in won't go out. I enter through the 'jockey door' to push the last three black Hebridean ewes out, slip over in the effluent so my blue Dickies boiler suit, which is the farmer's onesie, is soaked in ovine urine. Hebrideans have the devil's horns and it is with an evil grin that they bounce away.

After unloading these last sheep from the trailer, I begin a cadastral survey: I take stock of the local state of nature by having a nosey wander along the lane, which is thin and slick with red mud from tractors and field run-off. The Far East had a Silk Road; this region of Western farming has a Silt Lane.

A flint wind, no good for man or beast, cuts in from the east, so I hug the near hedge, which shivers naked and is no comfort at all. Magpies, those proofs of desolation, flap beside me. Magpies were not created, they were manufactured in some fantastical factory; after the nuclear winter, there will still be magpies clacking, still be magpies on dubious missions borne by mechanical wings.

There is a constant stream of outsize JCB Fastrac tractors, requiring me to walk on the verge. Only one thing outdoes the yellow brightness of the Fastracs: the green of winter wheat pumped by artificial fertilizer, which creeps across the land in a low neon vapour.

A pebble handful of wood pigeons is thrown by the wind across the sky; a fat tick of a grey squirrel, stuffed with acorn or mast, grips the trunk of an oak, and indolently watches me pass. In fifteen minutes of walking I've seen magpies, a squirrel and wood pigeons. My ears are overfull with the seashell noise of wind, and the diesel bass of £60,000 super-tractors.

Then, a comma in the loud sentence of my perambulating: three-quarters of a mile to the south, there is a farm that hosts a game shoot. 'Thorneycroft' says the wooden laneside sign in scooped-out letters. Think what you like about game shooting, this farm has biodiversity; there is ground cover and there is food. In a field of spiky maize stubble I count twenty red-legged partridges. Dumpy, clownish-bright, they are a warming sight; next to the maize spears is a patch of white millet with a thousand bouquets of seeds, and these twing and twang with goldfinches and chatty house sparrows. Five little yellowhammers fly in.

I walk back into the monotone scene, past Flinders, and along the lane to the north. A blackbird has been spread-flattened into the road, pressed into the ooze. A car coming round the bend fish-tails on the mud. The driver gives me a resigned, apologetic half-wave.

After a mile, I reach a dairy farm. The roadside

yard is clinically smart, as befits modern health and safety, but I cannot help but notice that around the back of one of the steel-framed barns is an overgrown paddock full of bits of scrap.

All farms used to have an untidy corner where machinery went to die, and where thistles and nettles grew. Intensive farming has all but done away with these little no-man's-land nature reserves; modern farms are as obsessively tidy as showroom Hygena kitchens. I walk on, and see in the entrance to a field that a new pond has been excavated, the raking scrapes from a digger's bucket visible in the clay.

Here, then, is a farm with a commitment to conservation. Exactly as this thought sparks across the synapses, there occurs the strangest of synchronicities: an evidential jack hare runs, with the rocking-chair gait peculiar to the species, down the lane towards me.

He stops, and glares with golden eyes. Hares have the chiselled head of horses, the legs of lurchers – and the eyes of lions; the ancient Chinese considered the animal so other-worldly they decided its ancestor lived in the moon.

The hare, now up on hind legs, cock-eared, continues to look at me, unblinking. The Middle English poem 'The Names of the Hare' gives seventy-two synonyms for hare; they include, with the accuracy that comes from accumulated communal knowledge, 'starer'.

My God, hares are large: this tawny, magnificent

27

creature must be two feet long; almost half the size again of its drab rabbit cousin.

For a minute, maybe, the hare and I see eye to eye; then a tractor-juggernaut thunders along to interrupt the moment; the hare bolts through a hole in the tangle of the hedge-bottom and into the field. I go to the gate to follow its progress, and there it is, loping effortlessly through turnips.

The hare is *lepus* from the Latin *levipes*, light foot, because of speed such as this. 'First catch your hare', began Hannah Glasse's famous recipe for jugged hare in *The Art of Cookery Made Plain and Easy*, 1747, which is no easy feat when hares run at 40mph. (Alas, the infamous cooking instructions were misquoted. She actually wrote 'first case your hare'; 'catch' was far more entertaining and stuck.) Hare is strong meat, made stronger by being cooked in its own blood; a freshly killed hare is prepared for jugging by removing its entrails and then hanging it in a larder by its hind legs, which causes the blood to accumulate in the chest cavity. I can still remember my grandparents' cold larder in the 1970s, with hares hanging for so long that bits dropped off as one brushed past. 'More flavour that way,' insisted Poppop, regarding his hang-'em-till-high method. Jugged hare has all but disappeared from our tables. A survey in 2012 found that hardly any British children knew the dish or, indeed, would wish to eat it.

The Romans, who may well have introduced the hare to Britain, were keen hare-eaters; Pliny the Elder advocated a diet of hare as a means of increasing sexual attractiveness. My grandfather had five daughters, so Pliny's postulation may be true. Then again, Pliny's other proposition concerning hares was almost entirely contradictory: he declared the animals hermaphrodite – a belief which eventually got worked into Christianity. Hares are a recurrent motif in British church architecture, standing for reproduction without loss of virginity.

In the turnip field, another hare slowly rises from the ground, stretches, then lies down to become a clod on the earth again.

Hares! A mile from Flinders. A long way. Too far for them to travel to colonize? I don't know.

Walking back to the Land Rover at Flinders I become conflicted. Some walled-over recess of superstition splits open. A hare across the path is unlucky. After a small tussle, sense v credulity, I decide that our journeys did not intersect.

But I say, just in case, 'Hare before, trouble behind: Change ye, Cross, and free me.'

FOR THREE DAYS when visiting Flinders, I pendulum-walk between the shoot and the dairy farm, but only add starlings, rooks, black rags of crows, pheasant, buzzard, blue tit and rabbit to my wildlife tally immediately around Flinders. Snoopy the Jack Russell finds nothing worth chasing.

I erect a bird table in Flinders field, about ten feet in from the top corner, and as I'm putting out seeds the absurdity of it hits me: I'm feeding birds with the cereal grains they would have obtained naturally by any sort of halfway wildlife-friendly farming regime. Modern farming leaves no 'gleanings' – leftovers – or weed seeds. For the month of January I decide that I'll spend fifteen minutes a day observing the birds on the table, and fifteen minutes noting the birds in Flinders, plus the adjoining twenty-acre winter wheatfield owned by the Ramsdale twins, or the Chemical Brothers as I have already mentally dubbed them.

The £25 bird table from B&Q is more than a litmus test for birds; I'm trying to seduce birds to a home, to a haven. Until I can get planting in the spring there is little else I can do to seduce birds to the field. To the same end I also suspend two galvanized pheasant hoppers just off the ground, so gamebirds can poke their beaks in but not rats their snouts.

Plough Monday, next after that Twelfth tide is past,
Bids out with the plough . . .

Thomas Tusser, *500 Points of Good Husbandry*, 1580

FOR A THOUSAND YEARS or more in England, Plough Monday, the first Monday after Twelfth Night, was considered the date to start ploughing. Actually, 'Plough Monday' is a misnomer because on that day ploughboys tended to play ye olde version of trick or treat, and the mouldboard on the plough rarely turned earth. Ploughmen led a dancing procession through the streets, dragging a gaudily decorated plough behind them. On arriving at a house, the ploughmen asked for bread, cheese and ale, or a contribution of money. If they were turned away, they ploughed a vengeful furrow or two in front of the house. Trick! In the 'cockpit' centre of arable England – Norfolk, Cambridgeshire, Huntingdonshire, Northamptonshire, Lincolnshire, Leicestershire, Nottinghamshire, Cheshire, Derbyshire and Yorkshire – ploughboys were more refined and put on a mummers' play or 'jag' for the inhabitants of the abode called upon. Treat! Typically the jag ended with a pleading song, such as:

> Good master and good mistress,
> As you sit around the fire,
> Remember us poor plough boys,
> Who plod through mud and mire.
> The mud is so very deep,
> The water is not clear,
> We'll thank you for a Christmas box,
> And drop of your best beer.

Invariably, the plough jags featured a crone who was 'thrashed' to death and brought to life again. (The thrashing motions exactly replicated the manner by which threshers used flails to beat seed out of its chaff coat.) The symbolism is obvious: away with the old corn spirit, in with the new. There were accompanying acrobatic dances; it was hoped that the crops would grow as high as the dancers could leap. If the winter was severe, the procession was boosted by threshers carrying their flails, reapers bearing their sickles, and carters with their long whips; the smith and miller joined in too, because the one sharpened the ploughshare, the main cutting blade of the plough, and the other ground the corn. The assembled peasants wished themselves a plentiful harvest from the sown corn – as medieval peasants called wheat and rye – and that God would speed the plough as soon as they began to break the ground.

The Romans would have understood the ceremony, the woad-tattooed Britons too. The Christian God was absent from these rituals of medieval Britain. Pagan habits died hard; Plough Monday was truly a ceremony of propitiation, a relic of a rite as old as the plough itself. In Ancient Greece, Demeter, goddess of grain and agriculture, had been placated with an offering of the first fruits at a feast called the Procrosia, 'Before the Ploughing'. The Romans in Britain gave oblation to Ceres.

Ever savvy, the Christian Church absorbed the Bronze Age fertility rite into its own ritual. By the Puritan era, Plough Monday had moved to become the altogether respectable Plough Sunday (traditionally held on the Sunday after Epiphany, the Sunday between 7 January and 13 January). In many churches a ceremonial or church plough was kept in the church, in front of the altar of the Ploughmen's Guild, which was lit with tapers of rush or wax paid for by the local husbandmen, in order to ensure success for their ploughing and subsequent labours throughout the year. Otherwise, a plough was brought into church and blessed so that the year's labour might prosper.

I can tell you what the service entailed, because in East Herefordshire, at St Andrew's at Hampton Bishop, we still had Plough Sunday as late as 1981. Following the choir and clergy, a farmer who was also a churchwarden (in this case tweed-jacketed, brogue-wearing Mr Jenkins: a sartorial rig that said, 'Despite our vocal burr, We. Are. Fucking. Gentry') led the procession of the plough up the aisle; behind him came three farmhands (John Johnson and his big-me sons) who manhandled a vintage Ransome plough up the nave. On reaching the chancel step the farmer formally stated to the vicar his reason for bringing the plough to church, offering the work of the countryside to the service of God. The old iron plough rested on the soft-blue carpet of the chancel, while we warbled hymns. The vicar, and quite a few members of the congregation, passed the service staring at Melanie Williams's embonpoint, as though she were a latter-day fertility goddess.

The devil, though, had all the best plough songs, which were generally not work songs, more guild anthems. Plough songs were sung down the inn or during jags, rather than when one traipsed behind a flatulent horse or behind the mobile muckspreader that is the ox.

The most famous plough songs are 'God Speed the Plough', 'The Painful Plough' and 'John Barleycorn'; in this last the personification of barley is not a woman, as in the jags, but John Barleycorn, who is attacked and made to suffer indignities and eventually death. These correspond roughly to the stages of barley growing, cultivation, and brewing or distilling in alcoholic beverages.

The Penguin Book of English Folk Songs says this about it:

> This ballad is rather a mystery. Is it an unusually coherent folklore survival of the ancient myth of the slain and the resurrected Corn-God, or is it the creation of an antiquarian revivalist, which has passed into popular currency and become 'folklorised'? Some have also compared it to the Christian transubstantiation, since his body is eaten as bread and drunk as beer.

'John Barleycorn' was printed in the reign of James I but is said to be much older. There are as many as two hundred variants, but here's one of the best:

> *There was three men came out of the west,*
> *Their fortunes for to try,*
> *And these three men made a solemn vow,*
> *John Barleycorn should die.*

They ploughed, they sowed, they harrowed him in,
Throwed clods upon his head,
And these three men made a solemn vow,
John Barleycorn was dead.
Then they let him lie for a very long time
Till the rain from heaven did fall,
Then little Sir John sprung up his head,
And soon amazed them all.
They let him stand till midsummer
Till he looked both pale and wan,
And little Sir John he growed a long, long beard
And so became a man.
They hired men with the scythes so sharp
To cut him off at the knee,
They rolled him and tied him by the waist,
And served him most barbarously.
They hired men with the sharp pitchforks
Who pricked him to the heart,
And the loader he served him worse than that,
For he bound him to the cart.
They wheeled him round and round the field
Till they came unto a barn,
And there they made a solemn mow of poor John Barleycorn.
They hired men with the crab-tree sticks
To cut him skin from bone,
And the miller he served him worse than that,
For he ground him between two stones.
Here's little Sir John in a nut-brown bowl,
And brandy in a glass;
And little Sir John in the nut-brown bowl
Proved the stronger man at last.

And the huntsman he can't hunt the fox,
Nor so loudly blow his horn,
And the tinker he can't mend kettles nor pots
Without a little of Barleycorn.

The 1782 version by the Scottish poet Robert Burns had a definite Gaelic twist, concluding:

Then let us toast John Barleycorn,
Each man a glass in hand;
And may his great posterity
Ne'er fail in old Scotland!

But when Burns wrote these words Plough Monday/ Sunday and the plough rituals had already started their precipitous fall. The Reformation had purged the guilds' ceremonials, while the historical changeover of farming from arable to pastoral – there was more money in fleece and meat – meant the toppling of the ploughman from his position of pre-eminence on the farm. Then came twentieth-century atheism, plus the switch to autumn ploughing. When ploughing begins in September, a ceremony for Wotan/God/Whomever to speed the plough in spring is redundant.

I'M SPENDING AN inordinate amount of time reading about ploughing since I can't do it in actuality. In the lost land of ancient ritual it may have been possible to plough in January; in the wet west of today's England the ground is too sodden to walk on, let alone traverse with a tractor and steel plough.

I have now looked up 'Flinders' in our bulky, blue-cloth-jacketed *Shorter Oxford English Dictionary*, because words are more real when they are in books:

> Flinders (fli.nderz) sb. pl. rarely sing. 1450 (prob of Scand. Origin; cf. Norw. Flindra thin chip or splinter.) Fragments, pieces, splinters. Chiefly in phrases, as to break or fly in (to flinders).

I am beginning to think that my dream of a florulent wheatfield is flying into 'flinders'.

It is still raining. When will the weather change? Such is the plowman's lament. One windy night I go over to Flinders to check on the sheep and moonlight winks at me from shattered puddles.

THINGS ARE WORSE than I thought. The sheep have now chowed down the kale for me to see half the surface of Flinders.

Wind rakes the dull earth. A single magpie loiters by the sheep trough, scavenging crumbs of leftover sheep nuts. I can't quite put my finger on what is wrong until I get to the wire stock fence by the paddock, where the field slopes away from the wind. The sheep are huddled together, backs to the gale.

My head is bowed, with the involuntary focus on the ground that comes from butting into the elements.

I'm looking at the bottom of the stock fence, criticizing the man or men who strung it (a farmer's default complaint) because the staples have been skimped on and so the wire has been loosened by sheep's rubbing on summer days. There are six energetic lines of mole hills running down the paddock I'm renting, under the fence, into Flinders – but then the little red volcanic explosions stop. Once into Flinders, the moles have reversed away.

There are no mole hills in Flinders field.

A mole signals worms as reliably as a canary down a coal mine signals gas.

With a dull foreboding to match the day, I get the spade out of the back of the Land Rover, then walk into the centre of Flinders and dig a hole a foot square, a foot deep into the thick, cold clay. I do it again, again, again. The sheep watch, bemused. Worms do, admittedly, go down into the earth in winter – but not like this. In each hole, I'm finding two worms. Tops. The worms are slow and purple with cold and I rebury them hurriedly.

The main part of the field is effectively dead. I dig a hole in the two-foot-wide grass margin around the field; this has a much better worm quotient, five or so worms per hole. I just have to persuade the worms, by some gentle husbandry, to migrate into the field proper.

It starts raining. Again. Even so, I cannot resist putting my face to the rain to look at the Chemical Brothers' wheatfield next door. I'm mesmerized by it; I keep rubber-necking it in the manner of gawpers passing a car crash. Around the edge there is a white bleach mark from the overuse of pesticides and herbicides, in the crop itself are alopecia patches of bareness. There are cracks in the earth sizeable enough to insert a hand into.

Every time I behold the Chemical Brothers' field I start reaching for similes. As barren as Mars. As dead as a dodo. Desert-like.

I then do something illegal, and push through a gap in the hedge and up over the fence to an especially bare, scorched-earth corner of the Chemical Brothers' field, and dig a hole. Actually, 'dig' is the wrong word. Despite the recent rain, the ground is so compacted I have to chip into it with the spade. I give up after six inches. There are no worms. The field is rutted eighteen inches deep where the tractor has struggled to drag the plough through the wormless soil.

Here's the thing: earthworms burrow underground by muscular contractions which alternately shorten and lengthen their bodies, as they seek and then expand crevices by force. A worm is a piston, aerating the soil and providing drainage. A worm, in coming to the surface and bringing down organic matter such as leaves, then eating it, increases soil fertility. Worm faeces (casts) are five times richer in available nitrogen, seven times richer in available phosphates, and eleven times richer in available potassium than the surrounding soil. Charles Darwin, who knew a thing or two about nature, was the first person to write a comprehensive book on the ecology of earthworms, namely *The Formation of Vegetable Mould through the Action of Worms*, 1881. As Darwin noted: 'The plough is one of the most ancient and most valuable of man's inventions; but long before he existed the land was in fact regularly ploughed by earthworms. It may be doubted whether there are many other animals which have played so important a part in the history of the world as have these lowly organised creatures.'

Forgotten Rule Number 1 of Farming: The more worms, the less the ploughing.

Darwin estimated that arable land contains up to 53,000 worms per acre. Fifty-three thousand worms per acre? Not in this field, not in this time. And not in Flinders either.

High irony: I am digging in the ancient Hundred of Wormelow, named for the River Worme.

The heavens open, and as I make my exit the Chemical Brothers' field follows me. The topsoil slips off in a pinky watery sheet, down into the shared ditch, adding more silt. Their twenty-acre field must be losing tons of soil per year. I cannot possibly do the maths, but soil erosion on some fields is such that losses of 47 tons per hectare have been recorded. In the last forty years the Earth has lost a third of its arable land due to soil erosion. The UK has maybe one hundred harvests left if we do not take better care of our soil. Winter wheat, as planted in the Chemical Brothers' field, exacerbates the problem because the land is tilled for the crop in time for the wet and wind of the year's end. (One Norwegian study in the 1990s estimated that autumn ploughing caused as much as 90 per cent more soil erosion than spring ploughing.)

Then I catch the smell of the ditch. It's the taint of sanitization. It's the ammonia whiff of nitrogen fertilizer.

Back in Flinders, I stick my hand into one of the spade holes in the field margin, and pull out a handful of this red earth that still has worm-life. As I walk back to the Land Rover I roll a mini-Earth with my fingers. The clay stains my hand.

On a whim, I decide to take my mini-Earth home, and put it under the kids' *National Geographic* microscope.

It is a journey into earth.

There is some initial cack-handedness; I put too much soil on the slide and see red mist when I peer down the eyepiece. Then I place a pinhead's worth of soil – literally – on the slide, add water and a glass cover.

You have not seen the world until you have seen a grain of earth under x600 magnification. There are easy-to-spot long threads of plantlike material, which I take to be fungi. Then, in this interior universe, come floating along jellyfish, cylindrical grey beasts so big I pull back my head, and wriggling transparent snakes.

There is life in the soil of Flinders.

And what soil: under the lens is a geological display case of rock shapes, glassy rectangles of silicates dominant. It is the silicates that give the red clay its shine when dug or ploughed.

Red is the colour of Herefordshire. When I lived away at university and reached Llangua on the way home, I knew I was back in Herefordshire, not because of the roadside sign with the county's name but because of the red of the ploughed fields beside it. They were sign enough.

I had, I confess, been doubting my vision of turning Flinders into a traditional wheatfield but there is something so miraculous about the hidden life of the soil that it feels negligent to doubt. How can a son of soil abandon soil?

That night I drive back out to Flinders to check the sheep. With the Milky Way above my head, the talismanic Plough configuration in the east, I walk the field in semi-blindness, because I want to *feel* the field under my feet. This earth, this clay from Old Red Sandstone, was formed 400 million years ago in this very place, when thick deposits of sand and mud accumulated in a basin and were stained red by oxidized iron minerals.

And I want to look at that other universe, the one above my head.

On air as clear as glass, I can hear the faint rumble of traffic on the A466, and the eerie, ecstatic cries of foxes mating in Three Acre Wood.

AT THE END OF January I tot up the visitors to my bird table. The square foot of board has attracted more avian species than the entire twenty-five acres of arable land surrounding it. Hardly scientific, but horribly illustrative.

The table has seen goldfinch, greater spotted woodpecker, house sparrow, wood pigeon, rook, jackdaw, chaffinch, hedge sparrow, blackbird, robin, song thrush, tree sparrow, blue tit, great tit, coal tit, redwing. (The hoppers have attracted pheasant.) There is volume as well as variety on this Piccadilly Circus of bird tables, particularly now that I have draped it with more seed feeders and suet holders.

The bird species in the Chemical Brothers' field and Flinders total, in one month, wood pigeon, pheasant, hedge sparrow, rook.

DRIVING TO FLINDERS to offload a hay rack I go past the shoot at Thorneycroft, where a dozen men suave in tweed, from knee-breeches to cap, are standing beside the road. They have expensive green wellingtons. Aigle. Barbour. Hunter. Le Chameau. The smooth-cheeked gentlemen are 'guns' up from the City. Slightly apart is a miscellaneous group of men and women in anoraks holding sticks with flags; these are the local 'beaters'.

I've got Edith in the Land Rover cab beside me; she looks out of the window at the other black Labradors, the ones waiting beside the guns. As we go on, she turns her head 180 degrees to continue staring at the guns and the beaters and the retrievers, clearly believing it a mistake not to stop.

She then stares disdainfully at my wellingtons. Underneath the mud and battering of four years of constant wear they were Aigles once, and new.

The problem for the shoot is that birds missed by the guns get dispersed across the countryside. I'm hoping, of course, that some red-legged partridges, in their diaspora, will come down to Flinders.

Half an hour later, the shotguns start up. It is the sound of money, welcome enough in one of the poorest counties of England.

I'm at Flinders to feed the sheep with ewe nuts: they flouncy-bottom run to the trough, a remnant of lambyness; then, when I am deemed too slow with my shoulder-sack of food, cat-tangle my legs.

Some left-over medieval mist hangs in the trees of the wood, and out of the white ages come the rooks, their caws as harsh as rust.

Clever birds, rooks.

I suppose the positive in having so few bird species to look upon is that one gives the old familiars more attention.

There are eighteen rooks' nests in the wood at Flinders, black blots in the bare superstructure of ash trees. Elms used to be the rook's favourite tree to build in, but when Dutch elm disease wiped out over 99 per cent of the elm trees in the 1970s, the rooks switched to oaks, sycamores and, as at Flinders, ash.

Naturally, the rooks are visiting the bird table in search of easy pickings. To prevent them from eating absolutely everything I have hung a seed hopper and peanut hopper on hooks; these devices are patented 'bad bird'-proof.

Or maybe not. If I were not watching this with my own eyes I would not believe it. Two rooks fly in. One unhooks the seed hopper so it falls down and spills its insides; the other rook, having watched this exercise in breaking and entering, unhooks the peanut dispenser so it too plunges to cascade its contents like a gaming machine with a winning line.

I suppose I should not be surprised. In a Cambridge University experiment five hungry rooks actually manufactured a hook from a piece of wire so they could hoist a small bucket of worms out of a tube. Rooks have an intelligence to rival that of chimpanzees.

Down on the ground, the Bonnie and Clyde rooks rub beaks with each other, presumably the corvid equivalent of the human gangsta's high five.

I leave the rooks their well-gotten gains, and instead drive to Countrywide in Hereford to buy a 25kg sack of bird food.

There is penitence involved here. I am feeling guilty for the rooks I have killed in the past.

CANDLEMAS, 2 FEBRUARY The day which, in the country-side, traditionally marked the end of the sexual abstinence practised since Advent; nobody down on the farm wanted heavily pregnant women in harvest time.

Great idle puddles lie across Flinders field. The sheep have almost finished the kale, and the last stalks stick out in the mist much as do the masts of sunken ships in Gothic seas. Across in the Chemical Brothers' field, the cold has caused the wheat to stall at the stage where it flops like green starfish; a cock pheasant makes the long march through the field, eyes fixed to the ground, but finds nothing.

Edith dashes into the hedge, and refuses to come to call. So I pound over and pull her out by her collar, but don't see the blackthorn switch, which catches me

under the eye. Blood drops down my face into my mouth.

In the early dusk, car lights flit on the lane. Traffic is transient; I walk bleeding across the eternal land, and then I understand. What does Herefordshire clay earth smell like? It smells like haemoglobin.

As I close the gate to the lane, a robin tipples in a little song. Does the robin know I admire him? I believe he does.

I've driven only ten yards when I see that a clump of lesser celandines has flowered on the roadside verge. Slowing down, I spot snowdrops tight under the hedge. The floral white bells are late in ringing this year. If the snowdrop is native to anywhere in Britain, it is here in the west. Usually, I look for snowdrops on my daughter's birthday, 19 January. Curious, though, they should appear on Candlemas, since it is by the name 'Candlemas Bells' that they used to be known to country people elsewhere in Britain.

Flowers on one side of the hedge, but not the other, the fieldside. The roadside verges of Britain are our greatest unacknowledged nature reserve.

14 FEBRUARY IN deceitful heat, a small clump of redwings fly north; they are less showy than fieldfares. As the thin fawn birds leave the scene, a female kestrel slips in to perch on the telegraph wire which runs across the field. Kestrels are dimorphic, meaning male and female look dissimilar. A male kestrel has a grey

head, the larger female a brown one. There is something saucy-postcard preposterous about a moustachioed male kestrel and his big wife, but love, especially on St Valentine's Day, is blind. She makes a *kee-kee-kee* courtship call, and he comes scythe-winged out of the sun to fly above her; he stoops, and she spirals coquettishly up before him, and they play kiss-chase all the way to the wood.

This is now the third time I have seen kestrels at Flinders.

IN THE NIGHT comes Jack Frost, the elemental serial killer of farmland birds. For morning after morning, Flinders is encased in minus-three white iron. In frost the birds of open farmland have it bad. The tree creeper can still find food in bark and crevices in the wood, but what hope rooks and starlings? The frost renders the ground far too hard for them to pickaxe for worms and grubs, which comprise the mainstay of their diet; unlike the majority of the corvid clan, crows, ravens and the ubiquitous magpie, the rook does not scavenge for carrion.

I am getting rather fond of my rook neighbours, and put out extra mealworm on the bird table. Despite this, on the fourth day of Siberian frost there is a dead rook in the field. A rat patters out from the hedge to look at it, then scurries back.

I know why. Under the fence closest to the wood, a fox has slunk and is standing, breath panting white, ears pricked forward, calculating risk. She can see me at the opposite corner of the field. She looks at me, looks at the black fallen bundle of rook. Then darts, grabs with her mouth and lopes away, each pad-fall sending up a small puff of crystals.

Frost has an antique charm, but after three days it wears thin. The world is old, and I feel it this day.

Buzzards in the valley over St Weonards' ridge mew pitifully as they circle the woods. If they come this way they will find small if beautiful pickings: there are the multicoloured corpses of two cock chaffinches on the waiting earth table. I seem to be losing birds, not gaining them.

OH, FEBRUARY. AFTER four days of frost, you do nothing but rain, and then you snow. The sheep have finished the kale, and so I move them into the grass paddock next door. Flinders itself is covered in a shroud an inch deep.

Snow is the white stuff of nature detecting: there are distinct marks of rabbit coming out of the wood across the paddock into Flinders; a fox has padded in their wake, then veered to the hoppers, where the beat marks of wings on snow tell a tale of a quick escape. From the laneside hedge into Flinders, a flurry of tiny paw marks goes up to the bird table, where shrews and mice have ventured for the overspill.

A mouse-shoal of house sparrows is on the table now, clamorous, then down on the ground, then flying to the hedge at some imagined danger. The house sparrow, once a bird of the town, is now almost exclusively a bird of farmland. *House* sparrow: a bird which has lived alongside us since the Stone Age has declined in London by over 90 per cent.

Despite the snow, the sparrows emanate a discernible lustiness. They always do. In Ancient Greece, sparrow eggs were sold as aphrodisiacs; for Chaucer in *The Canterbury Tales*, the sparrow was a synonym for concupiscence: 'As hot, he was, and lecherous as a sparrow.'

The numbers of the house sparrow are also down in the countryside, which is their last bastion. Granaries from which one would have seen a hundred sparrows emerge thirty years ago are nowadays bird-proofed for efficiency (read money), but also health.

In observing the state of Britain's farmland one becomes an accountant of miseries.

And yet . . . Snow falls steadily, making the world anew, burying the land. I'm in the paddock beside Flinders at the decline of day, feeding the sheep, the sole real figures in a snowstorm toy. Then a soft-winged barn owl, its back as golden as a storybook grail, drifts across Flinders.

I want to sing. I wish I could.

THE LAND THAWS, and standing at Flinders it occurs to me that every spring is a replay of Britain emerging from the last ice age. Nature is all circles and cycles.

ON THE LANE to Flinders there is a heron.

I kid you not, a grey heron standing in the middle of the flooded byway, fishing. He looks at me with a sense of ancient entitlement, and flies away with distinct bad grace as I edge close.

Even in a Land Rover, the floodwater on the road is 'interesting', as in the Chinese curse, 'May you live in interesting times.' At one point, the swirling brown flood bodily lifts the one-ton Land Rover and deposits it twenty yards backwards.

When I finally get to Flinders, the lower portion of the field is one extended puddle. The wind ploughs long furrows in the trapped water.

Heavy with melt-water and rain, the ditch has its own violent drama. Snagged across the pipe which leads into the next field is a red stole dropped by a giddy gentlewoman tripping home from a ball.

Except, of course, it is not. It is the corpse of a year-ling dog fox. The snows and rains of winter are hard on predator and prey alike.

Over the field steals a silent shadow, which passes within thirty feet of me. With its trademark fork tail and long, pterodactyl wings, the red kite is a bird unlike any other. The temperature drops by a degree or more. Hard pressed, the kite has abandoned its usual killing ground of the heathen hills to the west and come raiding on this arable land. Good luck with that, I think.

In my doubting mind, I think the world will always

be winter. But there is hope. In the roadside hedge dog's mercury is pushing up. Plants tell the seasons as definitely as the calendar. If it's dog's mercury it must be the cusp of spring. Surely?

THE ROOKS ARE squabbling in the wood, pulling their neighbours' nests apart, thieving material. It is this penchant for pick-nesting that has caused the bird to enter the lexicon as a robber. The phrase to rook, meaning to cheat, was everyday slang for the generations born before the Second World War. A rook is a crook. In the bad old days of Charles Dickens, a 'rookery' was as likely to be a teeming slum of London criminals as it was a colony of *Corvus frugilegus*.

The rook, the bird for whom no one had a good word. Somehow, even the rook's admirable habit of undeviating flight became 'as the crow flies'.

I'm looking at the rooks and wondering: why live communally if one disputes all the time? See, I'm doing it too, bad-mouthing rooks.

Rain pecks my cheeks as I leave Flinders.

28 FEBRUARY I walk into Flinders to pull out a hay rack; the ground is so sodden-soft it feels like be skin sloughing off the body of the planet. For every two steps I go forward, I slip one back.

The sky is rumpled, cotton-white, an unmade bed.

A shower of arrowheads shoots over my head and lands on the bird table to attack the mealworm: starlings, wearing winter's starry night on their feathers. One starling hops off and over the paddock fence on to the back of a sheep, in the manifestation of an ancient alliance; the starling picks off parasites (food), hence the local name of 'shepstarling'.

These are the first starlings I have seen at Flinders. The Department for Environment, Food and Rural Affairs has a list of twelve specialist farmland birds, birds whose life cycle ties them to agriculture. They are corn bunting, goldfinch, grey partridge, lapwing, linnet, skylark, starling, stock dove, tree sparrow, turtle dove, yellowhammer and whitethroat.

Actually, the list is too thin. Rooks, barn owls, pheasants, red-legged partridges, quail, corncrakes and cirl bunting are also specialist farmland birds.

Still, I feel a distinct sense of triumph in persuading a bird on Defra's list to Flinders. A pheasant, tail pressed to earth, head held aloft, *cok-coks*, then vibrates its wings in apparent celebration. A chaffinch trills merrily from the hedge, where ladders of cleavers are starting to climb, and the steely whips of ash are fecund with rabbit-nose buds.

These are more signs of spring. Winter is not invincible. The starlings are portents of something else. Hopefully, I am mending the broken heart of Flinders.

THE OLD ENGLISH name for March was *Hlyda*, meaning loud; and the wind is suitably flailing wet and wild across Flinders.

I can only trust in the truth of the ancient saw about the month: 'Comes in like a lion, and goes out like a lamb.'

By now it is early March; cock blackbirds are fighting on the lane, cock pheasants in the field. On the 5th I lie awake at home at 5.30am, the dawn chorus shimmering in my head. By the day, the dawn chorus increases in volume and variety as the birds sing their spring songs.

Still the ground is unsuitable for ploughing. I've taken all the ewes home from Flinders for lambing. All except Soo, a Juglet (black and white to non-sheepophiles) Shetland. Soo has been with us through change of farm, change of flock. When my daughter was a baby, Soo was a lamb herself. At some indefinable point Soo passed from being part of the flock to part of another family. Ours.

At fifteen years of age, Soo may well be the oldest ovine about to lamb in Britain. Soo has also managed to sprain her leg, meaning that in transport she will topple over. There is little option other than for me to spend the night at the Flinders paddock with her, sleeping in the Land Rover cab. So over I go in the Land Rover at 8pm on the 10th with a duvet to tuck around me, a dog to tuck around me, a thermos and the box of lambing kit.

I can't sleep. Jackdaws, the original dysfunctional family, are raving somewhere in the hilling land behind Flinders, because jackdaws are never quiet. Their call is *kya-kya-kya*, but squeakily high-pitched, an alarm rather than a lullaby. 'Jack' is an old term for small and for rogue, and the jackdaw is both. Every two hours I have to check Soo anyway. She is staring at the moon, a sure sign that the birth process is beginning.

It is freezing in the Land Rover cab, even under a duvet and a dog, so to get warm I take the reluctant Edith for a walk.

I like walking the lanes at night. At night you can pretend that the countryside still has birds, flowers, bees, animals. Frost clenches the ground; the flat top of the hedge is capped white.

You can still get miracles in the countryside at night. As we turn for home, I see that the moon has a giant glowing ring around it: a moon halo, caused by the moonlight refracting through ice crystals.

Yes, you can still get miracles. At about 4am Soo lambs in moonlight.

There is nothing uglier than a newborn lamb, and this one, like all the others, comes into the world in an alien pod of slime. (Or, if you were of a cruel frame of mind, you would say the membrane shroud anticipates exactly that future plastic life on the supermarket shelf.) The mother is mute throughout; sheep are prey animals, and try for silence when weak.

A tawny owl screams from the wood. It might be

March, the so-called first month of spring, but retreating winter has turned and stuck its talons into Herefordshire. I am shivering, despite the multiple gilets and coats which make me look like the Michelin man's fat brother.

Is the lamb alive? In this night scene in a paddock in darkest Herefordshire, the yellow encasing the lamb is the only colour.

Is the lamb alive? . . . Ah, yes, there is a tremble in the chest.

Will Soo do her usual magnificent maternal routine? Yes, she is licking the lamb, licking it. First the membrane off the face, then stimulating the lamb's body with her tongue.

The lamb is a black-and-white patchwork, because this is one of those magic nights when the designs of nature are perfectly coordinated. The lamb does a perceptible shake of the head. Then a snuffle. In dialect, the lamb is 'sharp' – vital. Within minutes it is up on its preposterous stilts.

Will the tottery lamb find the teats? The lamb head-butts all the wrong places . . . Ah, yes, the lamb has latched on. The tiny tail wiggles happily.

The lamb is beautiful. Because nothing is prettier, gentler-eyed than a new lamb. When the lamb has suckled, I pick it up and spray the navel in gentian violet to prevent joint ill. In my hands, I can feel the silken curls of its fleece. A girl. She will be kept and she will be called Moonlight.

My eyes are teary, but only, of course, because of the cold.

Soo – protective, annoyed – stamps her feet. I return the lamb to her and she leads it away across the paddock to join some shearlings I have left as company. They become shadows on the frost. Black. White.

All I need now is a south wind to kill the frost, then I can go a-ploughing.

The south wind does come. Two days later, garlic mustard starts to spire, long and tall, on the lane side of the hedge. Salad and green, its leaves really feel and look like spring.

CHAPTER II

John the Plowman

'The cut worm forgives the plough'

William Blake, 'Proverbs of Hell'

WE'RE IN THE KITCHEN. I'm rummaging through my 'man drawer' to find a garden thermometer to check the soil temperature, which for spring sowing of wheat should be 8 degrees centigrade. I'm regaling Penny with the story of how ploughmen used to tell whether the earth was warm enough to sow (they'd drop their trousers and sit on the ground; if the bare bottom could bear the earth it was warm enough). 'Shouldn't you do that?' she says. Stung both by the implied inauthenticity and the impugning of my hardiness, I shut the drawer with a dismissive flourish.

Luckily, I've managed to palm the thermometer.

Medieval peasants used to walk to different strips of land; I drive a Land Rover to seventy acres in five different lots scattered over fifteen miles, and I don't reach

Flinders field till late morning. The heater on the Land Rover is still broken and now the driver's window won't slide shut properly, so I've upgraded the canine hot-water bottle from Jack Russell. I've resorted to driving with Edith the Labrador flat on my lap.

Under a cold clear sky I make a slit in the earth with a spade, and insert the thermometer so that just the top is showing. Then repeat the process in a diagonal across the field, four times. Every time the thermometer comes up at 8 degrees or thereabouts, and the spade comes out dry enough for earth to fall off.

The steel spade stands in for the plough blade; if soil is sticking like porridge to the spade it will stick like porridge to the plough blade.

Temperature and dryness are good. For an extra check, I lay my forearm on the earth, and it is tolerably comfortable. Two rooks scull past with twigs in their mouths, and away in the wood their neighbours are still bickering over property rights. Three more rooks straggle past; the return of rooks to the rookery is as untidy as their twiggy nests. Rookeries are permanent establishments; pairs of rooks use the same nest year in, and year out. The rooks won't stop their hoarse cawing now until September.

Garlic mustard is growing, rooks are nesting. We are at the vernal equinox, when the day has equal hours of day and of night. It is time to plough.

Some time in the 1980s As I plough one spring morning, and the earth behind the tractor is turned over, to lie heavy, folded and sheeny in demi-hearted light, I glance across at the half-field already done. On the arithmetically precise ridges stand rows and rows of lapwings, spectators in the stalls to watch my labour.

It is chance that I look up and see what happens. The sparrowhawk flicks into the field, then swoops up, and with pure, distilled cunning aligns itself with the grey church steeple, so that it is almost lost in the background. In comes the sparrowhawk, wings set still, as though suspended by wire . . . a child's malevolent mobile. The lapwings see the killer only at the last moment. Up they shoot, twenty, thirty of them, safety in numbers. In the air they perform a synchronized tumble, showing off their white underside before flipping over to their iridescent green back to make a giant eye-blink in the sky. It is this synchronized belly-back flipover that gives the bird its name, 'Lapwing' being a corruption of 'lap-wink'.

The sparrowhawk makes an elementary mistake, swerving for one lapwing, then twisting for a closer one. There is contact – a collision, a few feathers fall like late snow – but the lapwing, though knocked back, readjusts mid-air and is away with the flock and over the tractor cab. Unable to slow sufficiently to turn and chase, the sparrowhawk settles for a face-saving glide into the ash in the hedge.

In the hawk's wake comes the wall of sound of thirty

60

lapwings calling in alarm. For a split second the engine noise of the Ford 7600 is obliterated, and all that can be heard in the world is *wah-wah-wah!* rising to a crescendo, as though I were locked in a vacuum with babies in distress.

In an oddity of nomenclature, the collective noun for lapwings is 'desert'. For the bird with the back of darkest, mouth-watering green? Perhaps the collective noun was a fantastic prophecy, because the lapwings deserted my village soon afterwards, as they did much of Britain.

I grew up with lapwings abundant and all around; they filled the skies and they covered the land like black-and-white confetti. In the last decade, my children have seen lapwings precisely twice. Once when a little off-course band took refuge from a blizzard on our hill farm, and once when we were travelling through Picardy. They, like other children, have become used to nature at the minimum, nature on the ration. And each generation will become used to less and less.

Geoffrey Chaucer in his poem the *Parlement of Foules* declared the lapwing to be 'fulle of trecherye', an acknowledgement that the bird's ability to lead a 'wounded' trail from the nest, to lure away hunters, was well known by the fourteenth century.

But, really, any treachery is ours. Due to the intensification of agriculture the population of farmland birds in the southwest of England has fallen by over 50 per cent since 1970.

It was good – and I was lucky – to have lived in the last of birdy England.

18 MARCH I haven't ploughed since I was a teenager, more than thirty years ago.

So, no pressure then.

We've a Ferguson TEF-20 diesel which we use as a yard-scraper, meaning literally that: with an attachment on its rear end, the tractor scrapes the mud and muck off the yard.

Even in its sixth decade, however, the Ferguson will plough. Nothing lasts for ever but a Ferguson, known to all in the countryside as 'The Little Grey Fergie', just might. To give you an idea of how robust the iconic Ferguson is, Sir Edmund Hillary chose the tractor as the means of crossing Antarctica. Fergusons were made at the Standard Motor Company plant in Coventry between 1946 and 1956; in 2003 a fifty-something TEF-20 was driven 3,176 miles around the coastline of Britain, the longest journey ever undertaken by tractor.

Ploughing four acres with The Little Grey Fergie will not be quick, neither will it be comfortable. There is no cab – but, then, I'm always banging on about the need to be in contact with the elements, so I'm just sucking that one up – and the seat is metal. On a spring.

If I was ploughing good earth with the 26-horse-power Ferguson, I reckon I could use a three-furrow

plough. On modern worm-unfriendly soil such as Flinders I am going to struggle with a ten-inch two-furrow plough.

I drive the Ferguson over to Flinders, the plough on the back, this fifteen miles at 10mph being the second-longest-ever journey on a Ferguson, and the second coldest. Or so it feels. When I arrive at Flinders my face is an ice-mask.

I'm still feeding the birds at Flinders, and while stomping around to regain some warmth put out seed on the bird table. I then do my now statutory fifteen minutes of observation.

The sparrows land first, followed by a robin and blue tit. The sparrows, as is their wont, are distracted by sex, and a gang of chirruping males chase a female into the swirling tangle of the hedge, where the blackthorn blossom is starting to clot the branches.

Two sparrows remain aloof. Finally, I see the chestnut cap. *Passer montanus* must be Britain's most overlooked bird; the tree sparrow was not recognized as a separate species from the house sparrow until 1713, and its Latin name, with its declaration of hills as the bird's habitat, is plain ignorant. The tree sparrow is a bird of open, low country – as long as there are decrepit trees for it to nest in. When the immemorial elm was removed, the tree sparrow, like the rook, lost its des res.

The tree sparrow lost something else in the seventies. According to my childhood bible, *The Reader's Digest Book of British Birds*: 'In autumn and winter they [tree sparrows] feed over stubble-fields and rick-yards.'

Remember over-winter stubble, those fields of straw spikes left by the combine harvester, which lasted from November until March, the length of the chill rule of Samhain, Celtic god of darkness? Perhaps you don't. Since the great switch in the 1970s from spring to autumn ploughing on the arable lands of Britain, over-winter stubble is hard to find. Only in one place is it still common: on Christmas cards. On Christmas cards, inside glitter-edged frames, pheasants and partridges still nostalgically perambulate the snow-crusted ruins of the cereal crop. But Yuletide art no longer imitates real countryside life. Only 3 per cent of arable land is left as winter stubble. So there you have another reason for the tree sparrow's 93 per cent decline since 1970.

Suddenly, the sun comes out as though it wishes to illuminate the Regency beauty spot of the tree sparrow's face.

Light changes everything. In the sun the rooks' eighteen nests in the wood go from blots on the treescape to black diamonds in an ash crown.

The wood is only two hundred yards from the field; I spend the afternoon kicking through leaves and staring up till my neck aches, checking for suitable holes in boles for tree sparrows to occupy. In a three-acre wood I find few, so in the end I phone up a friend to run up six bespoke nestboxes. Four will be fixed in a terrace row to a pollarded sycamore, and two maisonette-style to a larch six feet away. Tree sparrows mate for life.

There is a moment in the wood, when the rooks have all flown off, and it is as quiet as church.

It is Penny who asks the obvious question. 'Why *do* people plough?' I start to explain long-windedly, so she suggests I write it down.

No one has ever accused me of being an ace talker. Then again, I stuttered till I was thirteen so I have had less practice than most. ('V's were my dread; by a twist amusing to Satan my bus stop on the way home from school was The Vineyard; I either had to say – to the derision of everyone, from bus driver to back-seat boys – 'Winyard' or get off at the earlier stop and walk an extra half-mile. Funnier still, I was doing all this wearing a school cap. Never did me any harm . . . Actually it probably didn't: the forced exercise made me the fastest thirteen-year-old over 200m in School history).

To grow crops, the upper soil is broken to produce a seed bed. As early as Neolithic times it became apparent that the more the soil was pulverized the better the germination and crop quality. The earliest 'plough' was a crude pointed tree branch or antler, which was used to stir the soil surface and enable seed to be buried. The first recorded plough is found in pictographs on the monuments of Egypt, where it consists of a wooden wedge tipped with iron and fastened to a handle projecting backwards and a beam, pulled by oxen or by humans. To all intents and purposes, early ploughs were hoes dragged through the ground, breaking it but not inverting it. The Greeks developed ploughs fitted with wheels, which provided far greater control and manoeuvrability. Oak and elm were used for the beams, and iron for the shares, or cutting blades. Since iron was very precious, ploughshare metal was used and shaped into weapons in times of war – ploughshares into swords, as it were.

Roman and Anglo-Saxon ploughs were little different to these Greek ploughs, though the Saxons used a heavier plough with up to eight oxen. Saxon farmers fastened their draft animals to heavy ploughs by their horns or even tails to draw the implement through the soil; the latter practice continued in Ireland up to the seventeenth century, despite being outlawed, which allowed the judiciary to make a good living from fining miscreants. To reduce the number of turns, the Saxons made their fields long. The English 'furlong', one eighth of a mile, is derived from 'furrow-long'. An acre is the amount of land a man with two oxen could plough in a day, a square with sides just under 70 yards in length.

Horses were used from the eighteenth century; before then oxen were the main draft animals. It's no wonder that horses replaced oxen, even though a horse cost more to feed, and had less stamina. A horse could walk almost a mile an hour faster than an ox.

In the early eighteenth century a plough with a convex mouldboard of wood was introduced from the Netherlands into England. The mouldboard is what inverts the 'furrow slice' to bury weeds and vegetable matter while exposing the surface to the disintegrating forces of frost, air and wind. The furrow slices made by the mouldboard on the plough lie against each other at an angle of 45 degrees. The furrow also provides drainage.

The word 'plough' appears to derive from the Saxon *plou*, although the ultimate origin is unknown. Max Müller in *Lectures on the Science of Language* connects the word with the Indo-European 'to float'. The same word would be applied to a ship 'ploughing' through waves.

Ploughs are more complex than one might expect, and I spend an infuriating day with a spirit level, blocks of wood, tape measure and spanner, aligning everything so that the plough is 'straight' and will turn furrows of equal size and shape. Also the shares and mouldboard have to be glittering clean, so soil does not stick.

Without getting too nerdy technical, the Ferguson plough is fixed, meaning that the mouldboard of the plough 'throws' the furrow one way only. To prevent furrows piling up in the middle, I will need to plough in 'lands', a series of rectangles around which one drives in a spiral.

For the first land, I set two marker sticks parallel to each other at opposite ends of the field. I'm ploughing uphill, to offset the natural tendency of soil to slide down.

The first cut is everything; in previous centuries the head ploughman would 'open' the field, because if the line is wrong or wobbly, all consequent ploughing of the field goes out of kilter, and one can be left with the dread of dreads: 'short work', odd corner pieces which are difficult to plough.

Unlike in love, in ploughing the first cut is the shallowest, about three inches deep, and with only the rear mouldboard down. I've stuck a piece of yellow tape to the front centre of the tractor bonnet: a gun sight to help me aim at the marker pole. I start anxiously towards the marker, a bamboo pole with a white rag, which suddenly seems a continent away. A flag waving a sheet-sized welcome would have been more useful. In drawing the first furrow, I am Janus-faced; half the time I'm looking forward, half the time I'm swivelled back looking at the plough behind me.

Somewhere in the middle of the field I drift off course, leaving a small bend in my straight line.

Reverse, cut out the bump.

Second, third and fourth furrows are all deep, and tight against each other to create the central ridge or crown. My grandfather, Poppop, would have called this setting out of the field 'copping out'.

The hand of the past is heavy. There is a photograph of my grandfather ploughing, looking over the back of the tractor at the furrows. And looking at me.

Suddenly, I'm off. Steering one-handed, right hand resting on the mudguard, somewhat for the plough-man's nonchalant poise, somewhat for comfort since one ploughs on a permanent slant, because the wheels on one side of the tractor follow the previous furrow.

Hubris and falls, and all that. I start making elementary mistakes. At the end of the furrow you have to lift the plough as you turn the tractor round, and in sailing along I start turning the tractor too soon and get a sequence of curves.

From the seat, looking down at the plough, the sensation is less of earth being dug than of soil being spewed in rows from within the plough, in the way that icing is piped from a bag.

Up and down, round and round I go at 3mph, life at walking pace, creating a red sea. Birds fly down instantly, like seagulls after a trawler. The birds are the gaunt crowd, rooks, crows, jackdaws and starlings.

The rooks come so close they almost go under the plough in their eagerness to spot the first worm or grub. Once, when I turn on the headland, I see two buzzards perched on the ground.

The dead-mouse smell of diesel, the steady checka-checka of the Ferguson's engine, the birds eating in the wake of the plough – these are wonderful things. Less wonderful are the wet patches where the clay clogs the plough, requiring me to get off and clean it with a spade.

After three hours I stop for lunch. In a bid for ploughman credibility, I have a hunk of bread and block of Cheddar. *Pierce the Ploughman's Crede* (c.1394) mentions this, with the added ingredient of ale, as the traditional ploughman's meal. Alas, if you believed the 'Ploughman's Lunch' to have been a staple of England's inn and pub menu for millennia you would be disappointed. The 'Ploughman's Lunch' dates from the mid-1950s when the Cheese Bureau, in an admirable bit of salesupmanship, began promoting the meal to flog its dairy wares.

My memory of the real, in-the-field ploughman's lunch in the 1970s was that it consisted of a sandwich, nailed home by a No. 6 or Benson & Hedges cigarette. In the furrow in front of me is the bone-white stem of a clay pipe, suggesting that tobacco was the curse of the ploughing class in Victorian times.

Two male pheasants indulge in a passeggiata; they change colour, from mousey brown to lustrous bronze, as they emerge from the hedge-shade into the light of the field. Their splendour depends entirely on the capriciousness of the sun god.

Then: a skylark goes up in the Chemical Brothers' wheatfield . . . before giving up as though aware of the hopelessness of it all. Wait, I call, wait.

Back on the open deck of the Ferguson some lingering breath of winter eats into me, though I carry on because, as the Bible says: 'The sluggard will not plow by reason of the cold; therefore shall he beg in harvest, and have nothing.'

By mid-afternoon, with light and temperature fading, the metal of the mudguard is becoming ice-tacky on my fingers, and I resort to my MacWet equestrian gloves (other brands are available, though not as good). The gloves are warm, yet sensitive enough for me to feel the furrow wall through the steering wheel and to make adjustments to the plough. I've already got a blue topcoat on; there is no easy way to make a Ferguson's metal seat comfortable, and I know from previous experience that a cushion endlessly and irritatingly slips about. Sitting in a long coat is the smart option.

A top coat, MacWet gloves, scarf, Cordings tweed cap: in the mirror of my own mind I hardly recognize myself.

I'm lost in ploughing, in life at a contemplative 3mph as an infinite ribbon of furrow is squeezed out from the plough. I look out across the ploughed lands and see some black-and-white birds, and for an ecstatic second I think I see lapwings; then they rise and fly to the dove-cot at the farm across the road.

You have to have done ploughing to know the sensu-ousness of it. You take the plough to one field and by ploughing it you undress it to reveal the fresh flesh underneath. It is a godlike act too, because you are the little Creator, the maker of the landscape anew. You are Wotan.

Small wonder the ploughman was the king of farm labourers.

I will go with my father a-ploughing
To the green field by the sea
And the rooks and the crows and the sea-gulls
Will come following after me . . .

'I Will Go with My Father A-Ploughing', Seosamh
MacCathmhaoil (1879–1944)

I'M STILL PLOUGHING the next day. Mid-morning, when I get off the tractor to check the plough depth, there is a coin lying by the furrow wheel. Rubbed off, it is an 1899 Victorian penny. This is what we call 'dwarf money' or 'fairy money' in Herefordshire, old coins turned up by ploughing. I put it in my pocket for luck.

Back in the bouncing bucket seat of the Ferguson I find myself singing on a loop George Butterworth's set-ting of A. E. Housman's 'Is My Team Ploughing':

72

'Is my team ploughing,
That I was used to drive
And hear the harness jingle
When I was man alive?'

Ay, the horses trample,
The harness jingles now;
No change though you lie under
The land you used to plough . . .

Just so long as I don't start singing The Wurzels' 'Combine Harvester' when I get to cutting the wheat.

LIFE IMITATES ART. I'm contemplating 'sillion', Gerard Manley Hopkins's invented word for the shininess of ploughed earth in his poem 'The Windhover' when I see said bird in the paddock next to Flinders.

High there, how he rung upon the rein of a wimpling wing
In his ecstasy! then off, off forth on swing,
As a skate's heel sweeps smooth on a bow-bend: the hurl and
    gliding
Rebuffed the big wind. My heart in hiding
Stirred for a bird, – the achieve of, the mastery of the
    thing!

The kestrel hangs in the air, moves on in a small curve, anchors itself by an invisible chain again. For more than ten minutes the male kestrel, the tiercel, quarters the paddock on his mouse hunt.

I, of course, also hope to meet a mouse, so I can join with Robert Burns, who, in November 1785, upturned a mouse's nest while ploughing. He apologized in verse immortal:

> *I'm truly sorry Man's dominion*
> *Has broken Nature's social union,*
> *An' justifies that ill opinion*
> *Which makes thee startle*
> *At me, thy poor, earth-born companion,*
> *An' fellow-mortal! . . .*
>
> *Thou saw the fields laid bare an' wast,*
> *An' weary Winter comin fast,*
> *An' cozie here, beneath the blast,*
> *Thou thought to dwell,*
> *Till crash! the cruel coulter past*
> *Out thro' thy cell . . .*
>
> *But, Mousie, thou art no thy lane* [you aren't alone]
> *In proving foresight may be vain:*
> *The best laid schemes o' Mice an' Men*
> *Gang aft a-gley,*
> *An' lea'e us nought but grief an' pain,*
> *For promis'd joy!*

My plan to encounter a mouse, of course, goes awry. It was corn, though, that brought mice and men together since both feed on it. That was in 8,000 BC, when man began to cultivate the wild grasses.

THE ROOKS ARE still closest behind the plough, closer than the other birds, less scared, the first to get the invertebrate treasures conveniently turned up by the steel. When I stop for elevenses, they spread out and go over the ground again, this time probing with their own built-in facial mattocks.

Rob Pryor, the contractor from Llanwarne, drops by, largely for some tractor banter.

Modern tractors, such as Rob's New Holland T8, can cost £250,000 and are as technologically advanced as a supercar. Which is why Lamborghini make tractors.

'Come and see how warm my cab is, John,' says Rob, who is wearing his usual All Blacks rugby shirt, an affiliation derived from a summer shearing course down under when an agricultural student.

The New Holland's cab is a battery of red and yellow buttons, ergonomic knobs and computer screens. There is a Sidewinder armrest from which all the key controls can be operated. What Wind-up Rob really wants to show me is the leather seat: 'Comfy, that,' he says, before making an exaggeratedly pained face at the iron-hard seat of the Ferguson.

Then, a theatrical rubbing of hands: 'Ooh, bloody hell, John, it's parky out here, let me get back in.'

Rob drives off with a teasing roar from his 290-

horsepower engine. I get back on the £1,500 Ferguson and plough on regardless. I plough my own furrow.

Mid-afternoon it starts to rain. A black wall comes towards me, then wraps me in hissing water. I get off the tractor and run, sideways into the squall, for the slight shelter of the hedge, where I crouch and watch the furrows fill. It is unendurable. Only a cock pheasant, the tailless one, the loser, ventures out and saves the field from loneliness. He minutely scours every clod and clump, and must get his due reward because he pecks away metronomically.

For a moment May Hill, the southernmost point of my world, appears in view, and I run to the Ferguson and tie the tarpaulin over it, then dash to the Land Rover and head for home.

That night I google the T8 online brochure. Scrolling down I find:

FARMING LUXURY The T8 luxury pack has been designed for those of you who spend more time in your cab than out of it. The full leather steering wheel, leather seat and deep pile carpet are available on all T8 models.

*

Ploughland/Arable Farming Words
Balk – ridge between two furrows
Capper – crust formed on recently harrowed land by heavy rain

Casting – ploughing the headland in an anticlockwise direction

Cop – ridge formed by two opposing rows of ploughing

Costrel – a flask for carrying beer or cider into the field

Coulter – metal knife or disc which, on a plough, cuts the ground vertically

Crown – main ridge when ploughing by lands

Dallop – patch of ground among growing corn that the plough has missed (East Anglia)

Furrow – narrow trench or groove made in the ground by a plough

Gathering – ploughing the headland in a clockwise direction

Headland – strip of land at edge of field where machinery/horses are turned

Land – sub-division of a field, usually about 12–20 yards in width

Mouldboard – part of the plough that turns the earth ('mould') up and over to create a gap in the ground, a furrow

Plough-pan – compacted soil, the result of repeated ploughing

Reen – interval between ridges of a ploughed field

Share – wedge-shaped steel blade before or on the front of the mouldboard on a plough that cuts horizontally

Short work – tiny corner of land missed by plough

Sillion – the shining, curved face of recently ploughed earth

Strip lynchet – bank of earth that builds up on the downslope of an often ploughed field

Ucking – amount of land to be ploughed in a given time

Veering – Herefordshire dialect for a land, after the veering or turning of the plough

Warp – soil between two furrows (Sussex)

*

DRIVING TO FLINDERS next afternoon in the Land Rover I'm humming round a bend when I spot a covey of five red-legged partridges on the lane; a certain amount of slewing and unintentional off-roading on the verge follows, which the red-legged partridges take in their short stride. By and large, red-legs prefer to walk or run, and I am able to herd them two hundred yards down the lane before they dart under a gate.

These red-legs are only a third of a mile from Flinders.

Damp from yesterday's squall has got into the engine and the Ferguson is reluctant to start, so I tinker about with spanners and cloths and aerosols.

The Ferguson is parked by the roadside hedge, close to the metal gate, a fixture a male chaffinch believes to be his, and he keeps up the species' trademark 'raincall' from the hinge post; this, I feel, is somewhat taking the piss. The endlessly repetitive *drip drip* is a type of aural torture. I am, of course, aware that a bird whose nominal prefix is 'chaff' has more right to be in a wannabe wheatfield than most.

Even so, I hiss at the chaffinch that his folk names include 'pie-finch'. Just saying.

Then I think: the sheer commonness of the cock chaffinch has blinded us to his magnificent handsomeness. The cock on the gatepost is in full breeding plumage, and Farrow & Ball would kill for a paint the slate-blue hue of his crown.

I don't start ploughing until after two. The birds have waited for me somewhere in the sky: they come out of nowhere as soon as the first furrow shines. There are seagulls now, twenty black-headed gulls, which fall behind the plough closer than the rooks even. Behind me is a constant white line of feathers. The black-headed gulls scream and shit, and a slash of excrement hits the Ferguson's front bonnet so that a fish-rot smell streams into my face.

I'm trying to rub off the seagull excrement when the male kestrel swings in, and I see a proof of how successful my bird table has been in attracting avians. He centres his first hovering on the bird table, from which the small birds explode away as though a bomb was dropped among them. The kestrel, however, has no interest in them but in the mendicant mice, shrews and voles which pray for dropped seeds.

Kestrels are birds of habit, and the tiercel has added Flinders to his daily round.

So familiar has the kestrel become along the edge of motorways that its status as a bird of open land, as a farmland bird, is all but forgotten.

Hoverhawk, windsucker, fanner hawk, windhover are among the country names which identify the bird's distinctive hunting style, although the Anglo-Saxon 'windfucker' comes nearest to capturing the dominating demeanour. Windfucker could also be applied to a rakish man. George Chapman, the Elizabethan dramatist and translator (of Homer, celebratedly), wrote, 'There is a certaine enuious Windfucker, that houers vp and downe, laboriously ingrossing al the air with his luxurious ambition.' This windfucker was good for nothing, just as the hawk was useless to falconers: small and untrainable.

'Kestrel', meanwhile, is from the Old French *cresserelle*, meaning rattle, an attempted phonic rendering of the bird's *kee-kee-kee* cry. The Old English *mushafoc* is for mouse falcon, and the bird does hunt mice, though the short-tailed vole is the staple food. A good year for voles is a good year for kestrels.

The vole is a luckless creature which unwittingly marks its paths through the grass with urine; this reflects ultraviolet that's detectable by the kestrel.

Red-backed in the Sunday afternoon air, the kestrel has sex with the wind, before skimming five yards on lean wings, as though let go from rubber bands, to try its chances further out. The kestrel slips forward again, hovers again. His head is stone-still. According to the naturalist Brian Vesey-Fitzgerald, a surveyor once focused his theodolite on the head of a hovering kestrel and reported that it did not move more than a centimetre in twenty-eight seconds.

Kestrels do not hover quite; they fly forward as fast as the wind is coming towards them. It is the wind over the wings that keeps the bird airborne, gives it lift.

The wind hesitates, and the kestrel compensates by beating its wings hastily, much as humans losing balance flap the air.

Finally, the kestrel's diligence is rewarded. It drops, fans the air; drops again, fans, hangs. Then dives victoriously from bungalow height on to something in the grass close to the fence.

I have my own small triumph too. I have finished ploughing Flinders.

After ploughing, the regimentally straight furrows lie next to each other, so that the four acres is made up entirely of parallel lines. The whole field gleams in low afternoon sunshine, as does the tailless pheasant, who stays with me as I clean the plough. All the other plough birds fly home one by one.

And so must I.

It is night as I close the gate on the day, and I see for the first time in more than forty years that starlight is sharper, harsher than moon glow. The furrows, so smooth and rounded by day, are geometric Vs now. The field is a mathematician's fantastic continent of miniature, precisely regular hills and valleys.

THAT NIGHT, AT HOME, I get out my copy of A. G. Street's *Farmer's Glory*. Street was one of the farmer writers who bloomed in the thirties and forties, along with Adrian Bell, Henry Williamson and John Stewart Collis.

I suppose, after having ploughed Flinders, I feel I can be readmitted to the fraternity of ploughmen. Street was as accurate as he was adamant about the pleasure of ploughing:

> Tis true I am no physician, but I would suggest in all sincerity that three months steady ploughing would cure any man of a nervous breakdown, for ploughing is a mental tonic of great power. The ploughman is master of the situation . . . In itself it is all-sufficing and soul-satisfying. You English townsfolk, who sneer at Hodge [the farm labourer] plodding at the plough tail, do not realize that he pities you in that you cannot plough and have never known the joy of ploughing.

THE EARLIEST KNOWN version of the 'The Painful Plough', the anthem of plough songs, is an eight-page chapbook titled 'The Ploughman's Garland', printed at Darlington in 1774. The folklorist Cecil Sharp noted that 'The adjective "painful" is, of course, used in its original sense of taking pains, careful, industrious':

> *Come, all you jolly ploughmen, of courage stout and bold,*
> *That labour all the winter through stormy winds and cold,*
> *To clothe your fields with plenty, your barnyards to renew,*
> *To crown them with contentment, that hold the painful*
> *plough.*

The ploughman then explains 'no calling I despise /
For each man for a living upon his trade relies', yet
gardener, Solomon, merchant, all are dependent on
the painful plough which prepares the land for the
cultivation of food:

> I hope there's none offended at me singing this,
> For it was ne'er intended to be ta'en amiss;
> If you'd consider rightly you'd find I speak it true,
> All trades that I have mention'd depend upon the plough.

I LEAVE THE PLOUGHED field alone for two days, so the
top will 'haze', dry and crumble, in the March sun.

When I go back, there are hallucinations in the
hedge. On the verge, two primroses are out, and above
them perched in the hedge is a primrose-headed yellowhammer, singing his rhythmic song, sometimes
described as 'little-bread-and-no-cheese'. The Reverend
C. A. Johns, author of the Edwardian classic *British
Birds in Their Haunts*, suggested that, 'if the words "A
little bit of bread and no cheese" be chanted rapidly in
one note, descending at the word "cheese, chee-ese", the
performance, both in matter and style, will bear a close
resemblance to the bird's song'.

Fifteen miles southeast of Flinders, the people of
Gloucestershire thought that the bird's song sounded
more 'Pretty-pretty-creature', while the Scots went for
the morbid 'Deil-deil-deil-tak-ye', deil being devil.

Something about the yellowhammer seems to have irked the Calvinists north of the border, who decided that a bird so gaudy must be the devil's work, and sustained by his blood:

> The brock and the toad and the yellow yorling
> Tak a drap of the devil's blood ilka May Morning.

The bird was persecuted hard in the north of Scotland.

So confident is the dandy yellowhammer of my swooning that he allows me within a few yards before he flutters off to land further along the hedge. We repeat the manoeuvre, then he flies off. Permanently.

A COLD SNAP, a 'blackthorn winter', shuts on us, traps us. The overnight temperature plummets, and Flinders is crusty with frost. The cock pheasants creep to the feeders, lurk there; the thin hens come once or twice, but they are already sitting tight on eggs in the wood. The crystal boughs of blackthorn blossom shake when the blackbirds launch themselves off to fly to the ever-giving bird table. Starlings flying in from the south beat and glide, beat and glide, as though skimming the furrow-waves.

A buzzard slides overhead on brittle wings.

The frost silently infiltrates the microscopic interstices of the ploughed soil and then explodes them.

Lying down in the field at night, ear close to the ground, I can hear the Pleistocene crackle of nature breaking up the soil, tiny cave by tiny cave.

This is the ploughman's true music. Nature herself making the tilth, the fine soil needed for a seed bed.

*

Only a man harrowing clods
In a slow silent walk,
With an old horse that stumbles and nods
Half asleep as they stalk.

Only thin smoke without flame
From the heaps of couch-grass:
Yet this will go onward the same
Though Dynasties pass.

Yonder a maid and her wight
Come whispering by;
War's annals will fade into night
Ere their story die.

Thomas Hardy, 'In Time of "The Breaking of Nations"'

*

IT'S ONLY ME harrowing clods, behind Willow the Shetland pony.

I've worked with horses before, and we've used Willow to haul logs and dog-carts. He's too small to harrow – to break up the furrows with a raft of metal spikes – much more than half an acre. He is here for me, because to walk behind a horse and harrow is to bring one into accord with all the ages.

Although I mostly communicate with him through the long reins, I talk to him too, and sometimes I whistle low, melancholy strains, which is the plough-boy's way of simultaneously soothing and stimulating his fellow toiling beast.

In harrowing half an acre Willow and I walk five miles. No one except kings and clergy was fat in the time of the horse. A man ploughing one single acre could expect to walk as much as ten miles.

The high-tone jingle of the harness, the clinking of the harrow when it hits a stone, the working-oneness of man and beast, the breath of horse in the coldening afternoon air, the proud lift of hoof out of soil, the distant cawing of the rooks – these are things English and lost.

I am happy harrowing, an emotional state which may, according to scientists at the University of Bristol, be enhanced by soil itself. A specific soil bacterium, *Mycobacterium vaccae*, activates a set of serotonin-releasing neurons in the dorsal raphe nucleus of the brain, the same ones targeted by Prozac. You can get an effective dose of *Mycobacterium vaccae* by walking in the wild, or gardening.

Or walking over a ploughed field.

But the wind. Hardy forgot the harrowing wind. I

have to stoop into it, ancient and beggared. I had wondered idly since A Level Eng Lit what it would be like to bring Hardy's poem, a favourite, to reality. It is an affliction shared by other farmer-writers; John Stewart Collis, working the land in the Second World War as recounted in *The Worm Forgives the Plough*, subjected Hardy's poem to critique by enactment. Collis, who found it 'difficult ever to say anything against Hardy (except with regard to Tess)', concluded there was 'no excuse' for 'half asleep':

> From the road a number of agricultural jobs look remarkably quiet, serene, slow and easy; but if you stand beside the man in question you may find that he is putting out all his strength, is moving quite fast, and is in anything but a serene state of mind.

Not only is it impossible to walk with ease behind the harrow, since you are stumbling the whole time over the clods, but you can't see your work properly.

It *is* the man who stumbles; to walk across a ploughed field requires five times the effort needed to walk across an unploughed one.

Walking in the furrows made by the plough is to walk between waves. When the rooks come into the wood to roost (my equivalent of the factory horn signalling the end of the working day), I have sea-legs and cannot for half an hour or more walk properly on flat ground.

MOST OF THE REST of the harrowing is done by the Ferguson.

If my overheads in old-tech farming are low, the physical expenditure is high. In the late afternoon, I drive along to the dairy farm to gape at the hares as a sort of downtime treat. The light is declining, and a hare comes out of the field to jink on the edge of reason by coming up to examine me from about ten yards. After all, I could be predatory.

The black-tipped ears are stone-unmoving; only the twitch of her nostrils proves that blood moves within her. For a prey animal, the hare manages a persuasive aristocratic mien. In Alison Uttley's Little Grey Rabbit stories it is Hare who is aloof and authoritarian: '"Where's the milk, Grey Rabbit?" asked Hare. "We can't drink tea without milk."'

Edith, in the Land Rover's cab, sees the hare's impertinence and scrabbles at the window to be let out for the chase.

Oh, dear dog, you are now matronly, and have not the chance of a snowball in Hell of catching a long-legged hare that can do 40mph, yet a solid 10 for enthusiasm I think. The hare, nonetheless, detects her intent, and bounds back through its smeuse, its hole in the hedge. Hares are haplessly regular; they are easy to poach because they always exit and enter fields by the same routes.

On the drive home I take a detouring loop to the Norman church at Kilpeck.

I arrive at the church, a compact three-cell affair of nave, chancel and apse on an egg-shaped mound, in darkness descending, and park beside the blowsy old yew at the gate. The ruins of the adjacent castle are jagged teeth in the last light of the west.

A robin machine-gun *tisks* as I enter the churchyard shadows. When Nikolaus Pevsner visited Kilpeck in 1963 while compiling the Herefordshire volume of *The Buildings of England*, he declared it: 'One of the most perfect Norman village churches in England, small but extremely generously decorated, and also uncommonly well preserved'.

If anything, Pevsner underdid the hyperbole, since the church, built by masons from the Herefordshire

School of Romanesque Sculpture, is awash with elaborate decorations. The church and its carvings are constructed of Old Red (Devonian) sandstone, an outcrop of which begins a couple of hundred yards to the southeast, and are still crisp in my torchlight after exposure to almost nine centuries of Borderland weather.

The building's most famous feature is the south doorway, ornately carved with Celtic, Saxon and even Scandinavian (Viking) art, with a tympanum over the top. But I am here for ninety-one jutting stones, or corbels, which run around the entire church in a table under the roof. One corbel is world infamous: a Sheela-na-gig, an exhibitionist woman in a positively tantric sex position with legs apart and long arms passing behind the legs to hold open the labia of her enormous vagina.

Other corbels include two men fighting, green men, a warrior entwined, and animals galore, some real, some fantastical, all indicative of how closely humans and the natural world commingled in the twelfth century. There are muzzled bears, deer, rams, lions, pigs, lizards, fish, cats, horses, serpents, goats.

Walking around the exterior of the church, my torch flashing, I'm looking for one corbel in particular. I've just been reading Oliver Rackham's *The History of the Countryside*, in which he declares said corbel to be of a dog and rabbit, and the earliest representation of bunny in Britain.

But surely Rackham is wrong? In my upright beam that is no rabbit, not with those long ears and those long legs. The Normans venerated the hare, including it as one of the four Beasts of Venery, the animals considered by the nobility to be worthy of *la chasse*. (The others were the boar, the deer and the wolf.) Besides, the rabbit had no meaning in medieval British Christianity, whereas the hare had umpteen symbolisms attached to it, including the fear of God. The dog next to which the hare lies is the face of faithfulness. In the earthly paradise the wolf shall dwell with the lamb, and the dog shall lie down with the hare.

On the way out, in the absolute departing of the day, I stop to look at lichen-encrusted family graves. My mother's ancestors were here so long they saw all of life's occupational possibilities, from lords to labourers, though mostly they farmed. When Thomas Gray wrote his 'Elegy Written in a Country Churchyard' he had graves like theirs in mind. Like Gray, I wonder how:

Oft did the harvest to their sickle yield,
Their furrow oft the stubborn glebe has broke;
How jocund did they drive their team afield!
How bow'd the woods beneath their sturdy stroke!

In the ash trees beyond the graveyard there are redwings babbling under moonlight, as they stiffen the sinews for the long voyage back to Scandinavia. They fly off, slipping drunkenly left and right as do Friday-night revellers in city centres, before they find the north star. I walk straight and south to the Kilpeck Inn for a pint of beer, the ploughman's prerogative, and sorely missed from lunch in days past.

THE LAST HALF-ACRE of harrowing; the honour goes to Willow.

Minus 1. Puffing-Billy breath, from the horse, and from me; dawn lies bleeding on the hill; rough, coarse Brillo-frosted ground, which hits up under wellingtons as I walk behind the harrow to make tilth out of the furrows. Walking over the symmetrical furrows it occurs to me why farming led to civilization: it wasn't just about food surplus and storage; it was about order. Maths. Geometry. Regulation.

CHAPTER III

Sow the Fields and Scatter

*If you will look at a grain of wheat you will see
that it seems folded up: it has crossed its arms
and rolled itself up in a cloak, a fold of which
forms a groove, and so gone to sleep.*

Richard Jefferies, 'Walks in the
Wheat-fields', 1887

ACROSS THE LAND, THERE is ploughing and harrowing into the balmy evening, green fields being transfigured to red, most for maize which will become cow-rations. The sky is scarlet, and there is no join between Heaven and Earth; they are welded together seamlessly.

You can smell spring in the air, its saladness; winter is bleachy oxygen, and the low, ferrous stink of rain on clay.

A robin and a blackbird jump into the field from the west side. In the wood, long-tailed tits are wind chimes in ash trees still winter-naked. I am wondering what other birds will advantage themselves of my corn-field-in-waiting, when a cock pheasant, quite fearless, totters in through the gate from the lane and past me. Dressed in his mandarin robes, he appears to be a particularly pernickety government inspector.

He scratches at my harrowed earth, he approves, he stays for half an hour.

I used to despise pheasants, as alien invaders, fowl bred by man – a Norman interloper bird not properly wild. Yet, in this time without birds, he who escaped the gun and the fox, and who has so consistently brought soul to Flinders, is welcome.

Hello, brother.

WHEAT IS SOWN; everything else is 'planted'.

Today's wheat traces its origins back over ten thousand years to the einkorn and emmer wheats that grew wild in the Middle East. The domestication of these varieties for use in agriculture and their arrival in Britain six thousand years ago displaced hunter-gathering in favour of farming. Wheat is the gold stuff, the food foundation of Western civilization.

Wheats grown today look similar to the old varieties, but have been selected for higher yields and better disease resistance. In Roman times wheat could yield three tonnes to each hectare; now eight tonnes is normal, though that sort of tonnage usually requires artificial fertilizer and showers of herbicide.

Wheat has its own strange terminology of Hagberg Falling Numbers (assessment of sprouting damage), lodging (falling over) and PGR (plant growth regulator). Hard wheats (high protein and starchy gluten) are sold for the production of bread. Soft wheats (low protein and weak gluten) are sold for biscuits and other general flour uses, while lower-quality wheats are used in animal feed rations.

Each golden grain of wheat contains three main parts: the bran, endosperm and germ. Depending on how the wheat is milled, various types of flour will be produced. Wholemeal flour consists of the entire grain, brown flour has some of the bran and germ removed, while white flour consists of the endosperm almost exclusively.

Wheat can be sown in either the autumn or the spring, both sowing times being harvested in August. In the UK, autumn sowing dominates because the temperate climate allows the plant to grow through the winter and produce a higher yield than a spring-sown alternative. The climate has always been well suited to the production of wheat and as much as a thousand years before the Romans arrived, farmers were exporting surplus grain to Europe. Britain currently produces around 15 million tonnes of wheat each year and around 25 per cent of this is exported.

There are seven main varieties of spring wheat available for planting in the UK.

I do not get a choice, however, because no dealer

contacted wants to supply my relatively small order. Eventually, I have to go tweed cap in hand to Rob Pryor the contractor, who gets an associate to bulk up his order, and Rob will bring over my share on his Matbro low loader, and then drill it for me.

The wheat is Paragon, which in the parlance is a miller's Group 1 wheat, suited for breadmaking. It has, at 90cm, a relatively long stem by modern standards. Nowadays straw has relatively little value and it is mostly chopped up and spread by the combine. Historically, straw would have been carefully saved for use in thatching and as bedding or feed for animals.

In a secret I have not shared with anyone, I am intending to harvest my wheat in a way that preserves the straw as fodder.

I have had second thoughts about hiring Rob Pryor to drill the seed. His New Holland tractor is the size of an urban semi, and will compact the earth. (Modern ploughing is a very vicious circle; large tractors crush the ground, meaning that yet larger tractors need to be manufactured to drag a plough through the compacted soil, and so on and on, ever larger.) I'm trying to tread lightly. Also Rob's hi-tech seed drill sows the seed with absolute regularity, and the sowing rate may lead to wheat too dense for hares and ground birds to live in. I phone him up again. We talk over the possibilities, and it's Rob who suggests the solution: 'If you want it old school, do it old school. Sow it by hand. Can't be that difficult, can it, just like sowing a lawn.'

He has a point, not least because I have just extended the grass field margin in Flinders on the side it joins the paddock to three metres by hand-sowing a pasture mix. Hares like grass, they like a view, and they like a sprint track.

26 MARCH Down the lane to Flinders comes Rob's red Matbro, my wheat seed swinging in a one-ton builder's bag off the front loader.

Rob plops the bag down in the field entrance with a mechanical-balletic flourish, before getting out of the cab. He's smiling more sadistically than ever. 'Here you are, John, your keep-fit seed.'

He tells me, quite rightly, that I am going to have a problem with 'weeds' choking the wheat and depriving it of food. Everyone tells me, all the time.

My problem is that I want to see what is in the soil, what it harbours, what it has saved from extinction. I also want to see my own sown wildflowers grow. The only solution is a sowing rate for wheat that will crowd out most, but not all, weeds/wildflowers.

Given that the old adage for hand-sowing is 'One for Rook, and one for Crow / One to rot, and one to grow', I am going for a heavy sowing rate of 500 seeds per square yard, meaning about two thirds of a ton of seed for the field in total.

'And what are you going to do about all the weeds in the straw when it's baled? They'll be as wet as buggery . . .'

says Rob, heading back to his purring Matbro.

Rob is no fool, and he is the first and the only person to guess my Professor of Oxford cunning plan for harvesting. I am walking behind him when realization, the Eureka moment, hits. He stops dead, sticks his two index fingers up into the sky, swivels in his Dickies Dealer boots, and points the two fingers at me, gun-style.

'Gotcha.' He swears, switches to 'Pardon my French,' and we stand together under the spring sky discussing it.

I am ten years older than Rob, and because I came down from hills sought by city-flight Londoners in my Land Rover, my Aigles and my RP voice, he thought, and these are his words, I was a 'gentleman farmer from off'.

When I explain my plan, I say, 'Financially, I'll win overall'; he says, 'You mayn't lose.'

My plan? Because I was born in the 1960s I have seen wheat harvested by a reaper-binder. I have seen wheat in sheaves, I have seen wheat stacked in golden stooks.

Once, it was just once, but I saw it.

The thing about wheat in a sheaf is that, if stood for a fortnight in a field with fair weather, the green stuff in it will dry. More, because the wheat straw is not buckled and broken by the combine it makes better and more palatable food for livestock.

*

IT TURNS OUT there is an art to sowing seed by hand. I settle for the method of the medieval peasant (yes, he's back from *Meadowland*) in the *Très Riches Heures du Duc de Berry*, c.1412–16, cloth sheet slung around my shoulder, open at the front to form a bag. In the medieval illustration the peasant-sower looks demented. I soon learn why.

It's inordinately difficult to broadcast seed evenly. I start off, walking up and down, spinning golden grains out of my hand, but the grains either clump or shower too thinly. Also, the cloth bag can only carry sufficient to sow twenty or thirty square yards, requiring frequent trips back to the one-ton mother sack. After two hours, I'm at home and on eBay searching for a seed fiddle.

You can purchase anything legal on eBay, including an antique seed fiddle for a 'Buy It Now' £150.

Sold to the gentleman from Herefordshire, collected that evening from Sally Pryce in Birdlip, and I'm a-sowing again by 11am on a self-satisfied spring morning, which preens itself. Starlings make relieved *phe-ew, phe-ew* whistles from the telegraph wire.

The seed fiddle is well named: the wooden box has a bow which, when pushed backwards and forwards like a violin bow, turns a disc in the box or hopper; seeds fall on the disc and are spun out of the fiddle as an even rain of wheat.

In west Herefordshire, the seed fiddle is not such an antique; farmers on Garway Hill used seed fiddles up to 1967.

The only, ah, fiddly thing is that the box is small, so the trips to the grain sack are still frequent. I said that the field is four acres; as I prove in walking up and down, up and down, it is 4.26 acres. Soil constantly encumbers my wellingtons, requiring stops every half-hour to scrape off my club feet.

After two days I am done with sowing, and done in. Farming? F***ing hell.

And I haven't mentioned the birds yet.

In the illustration from the *Riches Heures*, magpies and crows are eating the seed. Unaccountably, the illustrator missed out the jackdaws, rooks, sparrows and wood pigeons, which have descended on Flinders field in Hitchcockian quantity. The jackdaws actually follow my every step, as if they were my grey shadow. In the early medieval period small boys with stones and slings were tasked as bird scarers, but when the Black Death killed off half the people of Britain landowners resorted to technology, making clappers of three pieces of wood joined together, with which device a single child could deter whole flocks of birds. The other technological development was to stuff a sack with straw and top it with a turnip or gourd into which were carved faces – a scarecrow. I don't have a small boy, I don't want an anti-social gas gun, so I settle for a scarecrow.

Britain's medieval farmers did not invent the scarecrow; the imitation human is as old as farming. Ancient Greek farmers made wooden carvings of Priapus, who, despite being the son of Dionysius and Aphrodite, was hardly a looker, and his outsize erection (as in 'priapic') added to his grotesqueness. One gauge of the historical usefulness of an object is the number of folk names for it. Local names for a scarecrow include:

Hodmedod – Berkshire
Murmet – Devon
Hay-man – England
Deadman – Herefordshire
Tattie Bogal – Skye
Bodach-rocais (lit. 'old man of the rooks') – Scotland
Mommet – Somerset/Yorkshire
Mawkin – Sussex/Northamptonshire
Malkin – Northamptonshire
Bwbach – Wales
Wayzgoose – Cornwall

My scarecrow is a fence post, with a crossbar for arms, dressed in an old mac, a pillowcase filled with wool for a head (facial features courtesy of a permanent marker), and a bobble hat knitted years ago for one of the children by a grandmother. The scarecrow fills me with terror; the wood pigeons merely find it a useful place to rest. The rooks look at it in the manner of window-shoppers since, with their untidy feathery thighs, they have something of Worzel Gummidge's style in their own attire.

I should have realized the likely result of my labour,

because didn't Pink Floyd sing in 'The Scarecrow' that the scarecrow stood with a bird on his hat while guarding barley?

THE TAILLESS PHEASANT comes back, and brings a friend, and they dig and dig. My fraternalism with pheasants becomes frayed.

I provide them with amusement as well as food, since I have still to sow the wildflower seeds which I've purchased from Naturescape:

200g corn marigold (approx. 12,000 seeds)
10g corn chamomile (approx. 40,000 seeds)
10g cornflower (approx. 1,000 seeds)
10g corn poppy (approx. 10,000 seeds)

How exactly does one broadcast wildflower seed naturally to achieve Ye Olde Cornfielde effect? In the end I mix up the seeds in a carrier bag, walk into the field, and just fling handfuls of seeds into the air as if I were a declaiming poet. The minuscule black poppy seeds fall to earth like stone, but the cornflower seeds, those miniature shaving brushes for gentlemen, kite away. The rest do middling. Around and around I wander, as happy as a sunbeam, throwing seeds to the wind.

But not enough seeds. So I reorder online more corn chamomile and corn marigold, plus twenty plugs of both. And then, overcome by impatience, drive to the garden centre in Ross, and buy ten packets each of Unwins Wildflower Corn Poppy and Unwins Wildflower Cornflower, which I start scattering down the four edges of the field to create borders, or 'margins' in farming-speak. I grant that such a broadcast method is unnatural. It does strike me, however, that it is something easy enough for any arable farmer to do without 'contaminating' the crop.

By the end of the week I have sown something like 200,000 seeds.

It still doesn't seem enough. I cannot remember seeing cornflower or corn marigold growing wild for years.

Jackdaws tumble around the sky, iron filings moved by a magnet, as I sow the last seeds – 10g corncockle. As late as 1952, *Flora of the British Isles* termed it 'common'; these days if it is found growing wild it is a news item, followed by a panic.

This ancient weed, which came here with Iron Age farmers or the corn supplies for Roman legions, is mildly toxic. The seventeenth-century herbalist Gerarde wrote of corncockle: 'What hurt it [cockle] doth among corne, the spoyle unto bread, as well as in colour, taste, and wholesomeness . . . is better knowne than desired.' The seeds, largest of any cornfield weed (about 3–5mm diameter), and not much smaller than cereal grains, contain a poisonous sapotoxin, githagenin. Ground into flour, corncockle seeds may have increased susceptibility to leprosy; 3g powdered cockle seed is sufficient to bring about mild intoxication in man; over 5g can be lethal. (Corncockle is a case of cure and kill, however; the seeds are anthelmintic, and can 'de-worm' guts.)

In medieval times, the lord of the manor would send out small boys to 'de-rogue' the corn of the corncockle. Rather than poisoning all and human sundry, not to mention my own livestock, I am planting the corncockle in four small corner clumps, marked with sticks, and will uproot the plants before they seed.

On 1 April, and how I wish it were another date, I finish the sowing, by lightly harrowing the entire field to cover the seed.

MY 'CONSERVATION WHEATFIELD' has generated a bit of local interest; people stop by the gate to watch me work. One woman, seeing me use the seed fiddle, asks if there are other 'oddball farmers'; a couple of local farmers are downright hostile because of the possibility of wildflower seeds being dispersed, but as I point out, they are spraying for 'weeds' anyway; most like the idea and blame the disappearance of wildflowers on the supermarkets because, as one farmer in his sixties says, 'You've got to exploit every inch these days just to make a buck.'

He is right. Every time one tells a lie a fairy dies. Every time one buys the lie of cheap food a flower or a bird dies.

On one subject all the passing visitors are agreed: the wildflowers in the wheat will make the crop commercially useless, largely because the wildflower seeds will get mixed in with the grain when combining, but also because the green weeds in with the wheat stalks when baled will cause rot. 'I've a cunning plan,' say I, more in bravado than belief.

I think it is one of these passers-by who is my benefactor . . .

I DON'T GO TO Flinders for nearly a fortnight, because we are lambing, so I miss the first green push of wheat through the earth, which feels a little like missing a child's first steps.

When I do pop by to check how the wheat is getting on, I arrive to see that someone has tied plastic feed sacks along the bottom of the gate to make a barrier. Getting out of the Land Rover, I think initially that this is some hostile action, the imposition of a cordon sanitaire to keep my weeds in. When I push open the gate, there is a stone on the ground holding down a Ginsters pasty wrapping, inside which there is a note. The note, in black biro, says 'HARES!'

There are also tyre tracks a little way in, where a car has reversed. More to the point, when I look carefully across Flinders there is a large clod of unmoving earth on the green baize. And another. Someone has gone to a lot of trouble to deliver me hares. When I walk around the field, I can see that my benefactor has even stopped up one holey corner with rabbit netting.

I think I know the mystery hare-deliverer, an agricultural contractor who had stopped to talk and had mentioned, in passing, that he netted for hares, and on some of his jobs he wished he could relocate hares rather than spraying them as they lay still in their forms, their grass dens. I have never asked him since whether he was the mystery man and he has never mentioned it.

But will the hares stay? A hare can be ten pounds, so the stock fence may be more effective at keeping hares in than rabbits out, and yet I also want to attract more hares in, so the field has to be at least semi-permeable.

Also, is 4.26 acres enough land for a pair of hares, is there enough cover? I open up the gate to the neighbouring paddock, so they have the run of its five acres as well.

I am too fearful that my hares may leave to be excited.

Yes, I already think of them as my hares.

I gently edge the Land Rover into the field, and sit in the cab and wait and watch. Hares are active by day, usually more active at night, hence their long reputation as disguised witches. The seventeenth-century witch trials disclosed the following incantation:

I sall goe intill ane haire
With sorrow, and sych, and meikle caire
And I sall goe in the Divellis nam
Ay whill I com home againe.

As with many animals sacred to older religions, medieval Christians changed the hare into an animal of ill omen, saying witches shape-shifted into hare form to suck cows dry. Sailors considered hares so unlucky they could not be mentioned at sea. And not just sailors; country folk refused to call the hare by its name, opting for synonyms such as:

The hare-kin,
Old Big-bum, Old Bouchart,
The hare-ling, the frisky one,
Old turpin, the fast traveller,
The way-beater, the white-spotted one,
The lurker in ditches, the filthy beast,
Old Wimount, the coward,
The slinker-away, the nibbler,
The one it's bad luck to meet, the white-livered,
The scutter, the fellow in the dew,
The grass nibbler, Old Goibert,
The one who doesn't go straight home, the traitor,
The friendless one, the cat of the wood . . .
The hare's mazes . . .
The dew-beater, the dew-hopper,
The sitter on its form, the hopper in the grass . . .

The stag of the cabbages, the cropper of herbage . . .
The animal that all men scorn.

In the magic of twilight, my hares sit up and multiply, until they are five in number.

Is it just the hares, or is it the diffuse, restful light? The field looks different this evening, less . . . exploited, more comfortable.

The landscape is an ever-changing scene. As I ponder the philosophy of ontology, a mere speck of a bird flutters up from the centre of the field as if aiming for the stars itself. Through the open window of the Land Rover there is the sound of cascading, chinking coins.

My heart soars to reach the bird. A skylark has found the home I made for it, a home it would not otherwise have had. The corn is born, and has a skylark as its herald.

I have hares, and I have a skylark.

In the early spring, when love-making is in full progress, the cornfields where the young green blades are just showing become the scene of the most amusing rivalry. Far as the eye can see across the ground it seems alive with larks – chasing each other to and fro, round and round, with excited calls, flying close to the surface, continually alighting, and springing up again. A gleam of sunshine and a warm south wind brings forth these merry antics. So like in general hue is the lark to the lumps of brown earth that even at a few paces it is difficult to distinguish her. Some seem always to remain in the meadows; but the majority frequent the arable land, and especially the cornfields on the slopes of the downs, where they may be found in such numbers as rival or perhaps exceed those of any other bird.

Richard Jefferies, *Wild Life in a Southern County*, 1879

I CAME LATE TO Richard Jefferies (1848–87). He was Victorian, and what did Victorian authors, outside of the adventure writers Haggard, Buchan and Conan Doyle, ever do for any reader in their teens or twenties?

Thus I avoided Jefferies until my mid-thirties, when I was rootling through the shelves of Richard Booth's bookshop in Hay-on-Wye and chanced on *Wild Life in a Southern County*. I expected a purple sentimentalist. I did not find one.

Jefferies is invariably catalogued as a 'nature writer', which splendidly misses his point. In his (rather mischievous) essay 'Nature and Books' (1887), he expounds on the pointlessness of writing about nature by positing a question: 'What is the colour of the dandelion? There are many dandelions: that which I mean flowers in May, when the meadow-grass has started and the hares are busy by daylight.'

Is the dandelion yellow? Gold? Orange? All three? It depends on the time of day, the propinquity of other things, because the dandelion is 'like a sponge, and adds to its own hue that which is passing, soaking it up'. Jefferies' target is not really nature writing but abstraction and scholasticism over experience. There is 'nothing in books' which measures up to seeing and touching a real dandelion.

A farmer's son, Jefferies always liked an agricultural allusion:

> Lo! Now the labour of Hercules when he set about bringing up Cerberus from below, and all the work done by Apollo in the years when he ground corn, are but a little matter compared with the attempt to master botany. Great minds have been at it these two thousand years, and yet we are still only nibbling at the edge of the leaf, as the ploughboys bite the young hawthorn in spring.

Jefferies wanted to know 'the soul of flowers'. In going to nature, he sought a relationship with it rather than an explanation of its workings. He wanted communion. As he wrote in his autobiography, *The Story of My Heart*: 'I want to be always in company with the sun, and sea, and earth. These, and the stars by night, are my natural companions.'

He became increasingly mystic, unwilling to engage with the triumphalist Victorian world of materialism. Never physically robust, a longtime sufferer from TB, he spent the last decades of his life living in London, a curious end-of-life berth for a Wiltshire countryman.

I think he had given up. There is another reason why the badge of 'nature writer' sits uneasy on Jefferies' coat, quite apart from his disinterest in science. He saw the human figures in the English rural landscape, as well as the animal ones. (Indeed, he thought the singular fault in Gilbert White's *Natural History and Antiquities of Selborne* was the failure to include the human specimens of the parish.) Jefferies' subject was the countryside, the shared space of Humans and Nature. The mechanization of agriculture was doing away with Hodge the farm labourer and depopulating the rural scene.

The countryside, as Jefferies had loved it, had gone.

I WAS WRONG about the skylark; it does not nest in the field of the rising corn. I guess that its song fails to travel far enough for a mate to hear it.

The fauna which does move into the field is the rabbit, which decides that my young wheat is nicer than anything in Mr McGregor's garden.

LEWIS CARROLL created the defining image of the mad March hare, as ridiculous and minatory, in the tea-party scene in *Alice's Adventures in Wonderland* (1865). The 'mad March hare will be much the most interesting', decides Alice, 'and perhaps as this is in May it won't be as raving as it was in March'.

I'm rather hoping that my hares will be quite, quite

mad as late as April. So the next morning I'm at Flinders by 5am, with six straw bales to make a V-shaped hide, with the open back against the hedge. I'm inside by 5.54am.

The hour after dawn: nature's happy hour. Few humans are about; even farmers are still finishing their mugs of tea, unwilling to leave the heat of the kitchen Rayburn. These are the moments when the world still belongs to the animals.

The dew binds the night to the inch-high wheat blades, so the field is a black sea. A robin sings in the ash. And then as the sun comes over the hedge it brings a hare alive. She is just in front of me. She sits up, twists her head; her bulging eyes glow with the sun's fire.

She washes her face in a flurry of paws. It is cold, and her breath steams.

The jill has barely finished her ablutions when the earth births another hare, then another. One bounds towards the jill, and leaps at her.

She stands, and they fight, up on their hind legs, miniature humans. It's the scrappy pugilism of the playground, doggy-paddle blows, although effective enough, because fur flies in slo-mo cotton puffs. One of the males is no stranger to mornings like this; there are rips in his ears, lit through by the light, where he has taken blows from feet or bites from teeth in pugilism past.

I held a hare once, when I was about sixteen and played First XV rugby so knew physicality. The hare, a doe, had tangled herself in disused sheep netting at the side of the muddy farm track. With no care for her rescuer, she kicked at me with her sinewy back legs. I could barely contain her wild strength, and when she was finally free she used those same lengthy limbs to run away in bounds I later measured at eight feet. I was bruised blue and green for weeks after. 'She was quite an armful then,' as my father so deftly put it.

What lasted longer than the pain was the velvet feel of her brown fur on my hands.

For centuries it was believed that the hares in boxing matches were males strutting their stuff for admiring females. Rather, the match is between an unreceptive female and an ardent male. An XY hare is a jack. So the boxing is between jack and jill. It's a battle of the sexes.

She fights him off and sprints away, a cacophony of long limbs. Another stand-up fight ensues, yet quicker than before, lasting only a second or two before he retires to join the council of watching males.

She washes her face again. One of the other males rushes at her in the pheromone-induced madness of April hares.

One of my hares is missing; then I see it, off by itself eating, alone. It does not join the games.

With the sun rising, the hares go to ground. The wheat is growing daily, but is not high enough to hide a hare, so they creep to the grassy field margin where they have scraped shallow depressions. If they have feasted well in the night on the wheat and the vegetation, for hares are strictly vegetarian, they will have no need to come out to feed in daylight. During the day hares produce soft faeces which they then eat, meaning that the food in these droppings is digested a second time, extracting more nutrition from it. The faeces also contain bacteria which help to break down other foods in the stomach.

The sun's rays catch the dewdrops on the wheat, which flicker promisingly red and blue. Like miniature flowers.

THE HEDGES IN APRIL: of the three hedges at Flinders the laneside hedge is the oldest; according to botanist

Dr Max Hooper's famous rule (age of a hedge in years = number of woody plant species in a thirty-yard stretch × 110) it is four hundred years old and was a blackthorn-hawthorn hedge originally, which has since been colonized by wild rose and elder.

The top of the hedge has been flailed flat; it has not so much been trimmed as committed. Like a crime. (A hedge should be an 'A' shape.) Flat tops enable the easy ingress of rain and predators. Even so, at four feet in depth the hedge is still a sanctuary of thorns and hide-outs hidden by snaking ivy.

The hawthorn of the hedge has broken into leaf, pale and baby-tender. Does the hen chaffinch operate by a visual stimulus? The wisdom of generations? Some unconscious internal calendar? However she does it, her timing is perfect: she has finished building her nest to coincide exactly with the emergence of leaf cover. If she wears a drudge's dress, she is nevertheless an artist. The nest, in a hawthorn fork, is exquisite – a neat, rounded bowl of green moss, yellow and grey lichens felted with wool and spiders' webs, and lined inside with hair and feathers. There are five eggs.

No field should be an island entire of itself. But still: I am stringing barbed wire along gaps at the bottom of the stock fence inside the hedge in an effort to keep foxes and badgers out. I will make small 'pop holes' for hares and hedgehogs to enter and exit at will, but the field will be a sanctuary. An artificial one, certainly.

It is no good saying we don't want to interfere with nature. We already have.

ONE NIGHT AT EASTER, I'm on my way back from egg-delivering, and stop by the field to see what I still ridiculously consider 'my' hares, although few creatures are more remote from the possibility of ownership.

There is no moon or starlight; I have to wait for car headlights to come down the lane and scan into the field.

The Celts believed that the hare was the goddess Eostre's favourite animal and she changed into a hare at Easter. Caught in the sweep of car lights, the hares are not goddesses: they are small witches, dancing.

TO SEE LIVING THINGS I sometimes go to the place of the dead: the churchyard of St Michael's in Dulas, Herefordshire. Or, if you will, God's Acre in God's Own County. I went yesterday. Everything about the churchyard is perfect, beginning with the setting in a verdant valley far from the madding crowd. Two over-grown redwoods form a living lychgate, their resin a natural incense. Obligingly, walking up the path before me was a particularly pious pheasant in full episcopal rig, though if he wished to officiate at Holy Communion he was a decade too late. The church itself is decommissioned; only communion with nature takes place at St Michael's. When the church was built in 1865 the makers simply enclosed a bit of local field for burials . . . Within the drystone-wall boundaries of St Michael's is preserved a relic of ancient, glorious, traditional English hay meadow.

It is like looking back in time, to the age before agricultural 'improvement'. The wild daffodils are now over, but the bluebells are coming out to join the primroses, the cowslips, the violets. Later, there will be greater butterfly orchids, betony, black knapweed, tormentil, yellow rattle, vetchling, burnet saxifrage, and twayblades, and quaking grass – and all beasts and bugs who love and live on them. I have seen slow worms there, six-spotted burnet moths too.

Across the lane is Dulas Court, which was once a retirement home for musicians, some of whom are buried in the churchyard. There could be no more fitting place, because birdsong pours into the churchyard of St Michael's. A blackcap trilled vespers for me last night as I left.

There is a kind of salvationist hope in a churchyard; a belief that God's Acre will somehow escape the destruction of nature that comes courtesy of chemically addicted agri-capitalism, politicians' vanity projects in railroad-building, and Persimmon's wet-mortar dreams. Forty thousand acres of farmland a year is lost to industry, housing and roads. But, really, God only knows.

*

FLINDERS: IN AN April shower I stand and look at the young wheat. I do not mind the showers because these will cause the wheat and wildflowers to grow. Without moisture now, the plants cannot push down roots to develop stalks in time for the summer's sun.

What I do mind is that the green wheat is under attack from long black slugs, whose slimy mode of travelling is facilitated by the wet on the ground.

Aside from 'Wormelow', this district of Herefordshire has another name. 'Urchenwfled' or 'Archenfield', meaning the 'land of the hedgehog'. I could do with some slug-slurping hedgehogs, but they have been tolled by pesticides, and before that by persecution. The church accounts of Orcop over the hill show that in 1746 money was paid for four 'urchins'. In the past it was common for churchwardens to pay for the killing of vermin, and the hedgehog was considered vermin because it supposedly suckled cows' udders. Like hares.

It was not money well spent, I reflect ruefully.

My new friends the rooks are doing their pest-control best, however. There are twenty in the field, and they have not set a watchman as they sometimes do.

Perhaps I am their sentinel? Perhaps I have become adopted by the rooks? We have become used to each other now.

The rooks feed methodically, with silent purposefulness. In this spring sunshine their glossy plumage, as they turn and bend, catches the light and for a second they are surrounded in auras of silver.

There is the almighty clang of iron on concrete from the farm across the lane. The rooks look up, glance at me, then sail dilatorily to the far corner, and start their work again.

I've brought the binoculars with me, so I can see what they are eating. The rooks are destroying slugs, grubs and wireworms. They do much more good than harm.

Later that day: I am driving past Flinders and I see the rooks returning home, tired in flight, visibly clamorous. The wood closes over them in sanctuary.

I WENT TO DYMOCK in the Vale of Leadon, on the Gloucestershire–Herefordshire border, yesterday.

I spent part of my childhood ten miles away and Poppop and Grandma once farmed at The Priors on the Woolhope hills that shadow it. The farm never had adequate water, so my mother and her sisters – and this is true – had to go down to the public pump in the village, fill galvanized buckets and carry them back on a wooden bar across their shoulders. A yoke, by any other name. This is the late 1930s, and by Herefordshire standards of the time the Amos girls were distinctly posh. We are talking girls with names such as Josephine, Madeleine, Daphne, white dresses for confirmation, a cousin who's an RA portraitist. They also received dental treatment, admittedly done by Mr Balding the vet after he had attended the livestock, yet dentistry nonetheless.

Herefordshire is the forgotten county of England, including by the Electricity Board.

1969. Woodstock/Man lands on the Moon/The Beatles play their last concert. Oh, and west Herefordshire is put on the national grid.

I digress. Dymock is how you think of England: rolling meadows, hangers of trees, meandering brooks, red telephone boxes, deep lanes, thatched black-and-white cottages, and with some eccentric bumps around the horizon.

One such protuberance is May Hill, which is shaped awfully like a breast, the mammary effect enhanced by the grove of firs on top that approximate a nipple. The hill is one of the geolocators of south Herefordshire, always peeping through gaps in hedges, down the Vs of intersecting valleys, at the end of woodland rides. You know where you are when you see May Hill. It is our South Pole.

I love May Hill. Even as late as the early 1980s when I was a teenager it was Hippie Central. My friends and I would go there in a buzzing swarm of underpowered Honda 125s and Honda C45s – nice boys' motorcycles. (I think the motorcycles must have been a localized Herefordshire contagion: none of us, and we spanned council house to country house, had a car.) Midsummer was best, with a giant Crazy World of Arthur Brown

'Fire!' on the hill organized by flower people left stranded from the 1960s.

I wish I had known then what I know now. May Hill is where Edward Thomas and Robert Frost ambled on their 'talks-walking' when they were part of the circle of 'Dymock Poets', the commune founded by Lascelles Abercrombie. What united these poets, aside from a Romantic-worthy love of nature, was a pared-down verse style and a focus on the everyday, rather than the epic beloved of Victorians. They were 'the Georgians', new-style poets for the new king, George V. The American Robert Frost was a fulcrum, Rupert Brooke a distant satellite, and Edward Thomas in between as a moon. He spent weeks at a time at Dymock over the years 1914–15, and Frost's globally celebrated poem 'The Road Not Taken' is about the Thomas–Frost walk-talks; and as much as it concerns pure existentialism, the poem reflects Thomas's terrible personal havering when asked in which direction they should perambulate. Amused at Thomas's inability, Frost would chide him, 'No matter which road you take, you'll always sigh, and wish you'd taken another.'

So: yesterday I parked on the verge near Little Iddens, where Frost lived, and walked along the lane, following some of what is now the 'Poets' Circle Walk'. I do this periodically, as an act of respect to Thomas because, really, he sacrificed his life for the landscape of the Herefordshire–Gloucestershire border. He died as

Lieutenant Edward Thomas, Royal Garrison Artillery, at Arras in 1917.

When war was announced on 4 August 1914, Thomas and Frost were sitting on an orchard stile near Little Iddens. Thomas had planned to go and live in the USA with Frost but now Britain was at war. Thomas was no jingo; indeed, he refused to hate Germans or grow 'hot' with patriotic love for Englishmen, and once declared that his real country-men were the birds. He was also thirty-six, and exempt from military service.

So why did he volunteer to serve? He felt that his love for the landscape imposed a duty of protection. In walking with Frost around Dymock and Much Marcle and May Hill he saw countryside worth fighting for:

In April here I had heard, among apple trees in flower, not the first cuckoo but the first abundance of day-long-calling cuckoos; here the first nightingale's song, though too far-off and intermittently, twitched away by gusty night winds; here I found the earliest may-blossom which on May day, while I still lingered, began to dapple the hedges thickly, and no rain fell, yet the land was sweet. Here I had the consummation of Midsummer, the weather radiant and fresh, yet hot and rainless, the white and the pink wild roses, the growing bracken, the last and best of the songs, blackbird's, blackcap's. Now it was August, and again no rain fell for many days; the harvest was a good one, and after standing in the long sun was gathered in and put up in ricks in the sun, to the contentment of men and rooks. All day the rooks in the wheat-fields were cawing a deep sweet caw, in alternating choirs or all together, almost like sheep bleating, contentedly, on until late evening. The sun shone, always warm, from skies sometimes cloudless, sometimes displaying the full pomp of white moving mountains, sometimes almost entirely shrouded in dull sulphurous threats, but vain ones . . .

Then one evening the new moon made a difference. It was the end of a wet day; at least, it had begun wet, had turned warm and muggy, and at last fine but still cloudy. The sky was banded with rough masses in the north-west, but the moon, a stout orange crescent, hung free of cloud near the horizon. At one stroke, I thought, like many people, what things that same new moon sees eastwards about the Meuse in France. Of those who could see it there, not blinded by smoke, pain, or excitement, how many saw it and heeded? I was deluged, in a second stroke, by another thought, or something that overpowered thought. All I can tell is, it seemed to me that either I had never loved England, or I had loved it foolishly, aesthetically, like a slave, not having realized that it was not mine unless I were willing and prepared to die rather than leave it as Belgian women and old men and children had left their country. Something I had omitted. Something, I felt, had to be done before I could look again composedly at English landscape, at the elms and the poplars about the houses, at the purple-headed wood-betony, with two pairs of dark leaves on a stiff stem, who stood sentinel among the grasses or bracken by hedge-side or wood's edge.

On 19 July 1915 Thomas reported to 17 Duke's Road, London, to be attested Private 4229 in 28/ The London Regiment (Artists Rifles). Edward Thomas, one might say, went to fight for King and Countryside.

Edward Thomas volunteered not once but twice. Realizing that map-reading in the Artists Rifles was not sufficient, he applied for officer training. Again the countryside was the prompt. Thomas himself is surely the narrator in 'As the Team's Head-Brass'; the ploughman asks bystander Thomas the key question 'Have you been out?' Thomas can only answer 'No', and the negative is inadequate as Thomas watches the team ploughing the soil of England, his England:

> As the team's head-brass flashed out on the turn
> The lovers disappeared into the wood.
> I sat among the boughs of the fallen elm
> That strewed an angle of the fallow, and
> Watched the plough narrowing a yellow square
> Of charlock. Every time the horses turned
> Instead of treading me down, the ploughman leaned
> Upon the handles to say or ask a word,

About the weather, next about the war.
Scraping the share he faced towards the wood,
And screwed along the furrow till the brass flashed
Once more.
     The blizzard felled the elm whose crest
I sat in, by a woodpecker's round hole,
The ploughman said. 'When will they take it away?'
'When the war's over.' So the talk began –
One minute and an interval of ten,
A minute more and the same interval.
'Have you been out?' 'No.' 'And don't want
to, perhaps?'
'If I could only come back again, I should.
I could spare an arm. I shouldn't want to lose
A leg. If I should lose my head, why, so,
I should want nothing more . . . Have many gone
From here?' 'Yes.' 'Many lost?' 'Yes, a good few.
Only two teams work on the farm this year.
One of my mates is dead. The second day
In France they killed him. It was back in March,
The very night of the blizzard, too. Now if
He had stayed here we should have moved the tree.'
'And I should not have sat here. Everything
Would have been different. For it would have been
Another world.' 'Ay, and a better, though
If we could see all all might seem good.' Then
The lovers came out of the wood again:
The horses started and for the last time
I watched the clods crumble and topple over
After the ploughshare and the stumbling team.

Thomas was commissioned a subaltern in the Royal Garrison Artillery on 23 November 1916. He 'went out' to France in January of the following year.

Thomas was in good company in fighting for the land. Three of my relatives died in the Great War, two of them farmer's sons from Ocle Pychard. They knew the bounteous wildlife and flowers of Edwardian Herefordshire; they would, when they looked at the countryside around them, have also seen the work of their forefathers. Theirs was the beauty; some 75 per cent of Edwardian Britain was farmland, the result of agri-*culture* over centuries. Our landscape is man-made from nature. The countryside, to borrow the phrase of Poet Laureate John Masefield, is the 'past speaking dear'.

Men and women thought too, in the Second World War, that Britain's flora and fauna was a just cause. As James Fisher wrote in his preface to *Watching Birds*, 1940: 'Some people might consider an apology necessary for the appearance of a book about birds at a time when Britain is fighting for its own and many other lives. I make no such apology. Birds are part of the heritage we are fighting for.'

If Thomas had been with me yesterday he would have wept a little. There were some pleasing flashes of gold wild daffodils in the copse, but hedges galore have been grubbed out since 1914 to make bigger fields. Too much grass glowed with nitrogen-induced verdancy. A distant silver pool turned out to be polytunnels.

But it is still countryside worth fighting for.

*

I'VE SPENT HOURS scanning Flinders with binoculars, and have come to the dread conclusion. One of my hares, a jack, is missing.

20 APRIL The first swallows swoop over Flinders. There is something wholly uplifting in seeing the swallow, who, twice a year, undertakes the perilous journey to and from Africa, and whose life is an endless summer.

In the field margin under the laneside hedge, which is rapidly gaining its hawthorn leaf-coat, there are splatters of dandelions, drops of sun which fell to earth.

Curious how winter causes amnesia, and how in the spring things one has seen a hundred times before become startlingly novel. A bumblebee scours the hedge-bottom for a mouse hole in which to found a new dynasty. The queen is the only survivor from last year's colony; her transparent wings are a prism refracting light.

There are enough trees in leaf in the wood for me to wonder, and I have wondered this for a while: can one tell trees apart from the sound of their leaves in the wind? A sparrowhawk cuts around the edge of the wood, a rotating blade, chopping this way and that.

The wheat is four inches high, tall enough to tremor in the olid wind, throat-catchingly thick with dry chicken manure and yesterday's chemical dousing of the maize field across the lane.

Perhaps it is me who needs the cordon sanitaire, from my chemical neighbours, rather than my chemical neighbours needing a barrier against my 'weeds'. The first wildflowers in my personal ploughland are in bloom. And I did not plant them. They are scarlet pimpernel, and common field speedwell, both delicate bejewelled creepers over ground, the one red, the other blue. Their seed has been harboured safe in the earth for years; common field speedwell can germinate after twenty years. A wildflower to 'speed you well', the speedwell is as common on roadside verges as it is in arable fields, and travellers in years gone by sewed the flower into the lining of their coats as a charm. Those on metaphorical journeys into farmland admire it as well.

My own planted wildflowers are bursting through the earth, in green buttons and brooches.

There are also docks, thistles and nettles, and these I set about with a hoe because they are 'injurious', take-over-the-world plants. The kestrel comes by on his round; he goes about his work in the field, I about mine. As he cuts overhead I see he does bear a flying resemblance to a cuckoo, as folklore maintains; people once assumed kestrel and cuckoo were the same bird, the mysterious disappearing cuckoo turning into the form of a kestrel during the winter.

The hares have discovered the open gate to the

paddock, so have the run of ten acres. They are there now, lying low in their grass dens.

I rest for tepid tea from a thermos, under a cage of rose briar, which is snagged with bits of fleece; I start twisting them instinctively, the way the first humans must have done to make wool.

One of the hares comes out from its form, stretches slowly, then wriggles back in, hind end first.

Edith goes for a wander, and it is she who finds the missing hare; she drops on all fours, nose twitching, a compass needle for game. Sure enough, halfway down the ditch there are the hollow remains of a hare. Maggots crawl out through the eye sockets.

Although the day finishes cold, bats come out from the wood. The pipistrelles fly so close I can feel the chill swish of their black-leather wings. Bats are hardier than one might suppose, and can be kept in a refrigerator, given sufficient air.

There is not a day of my life that is not improved by seeing red-legged partridges. And, as I drive out of Flinders, there they are. A pair caught square in the Land Rover's headlights.

CHAPTER IV

The Golden Sea

Each morning now the weeders meet
To cut the thistle from the wheat
And ruin in the sunny hours
Full many wild weeds of their flowers
Corn poppys that in crimson dwell
Calld 'head-achs' from their sickly smell
And carlock yellow as the sun
That o'er the may fields thickly run

John Clare, from *The Shepherd's*
Calendar, 1827

THE YOUNG WHEAT SHIMMIES with spider webs. Interference on a TV screen.

At the grain hoppers are two red-legged partridge, one of which interrupts feeding to announce his arrival in the neighbourhood with a 'rallying call', sometimes related as *go-chak* or *sker-chak*, but easily mistaken for a steam-engine building up speed.

The beginning of May is late for a red-leg to be

claiming territory; I guess he and his hen have been moved on by predators or by a farming method which failed to appeal. In the morning sun the gaudy birds' legs are as scarlet as a femme fatale's lipstick. Red-legs were introduced into Britain as a gamebird by King Charles II, coming originally from southern France.

While the red-leg partridge has suffered decline in numbers, it has not dropped off the ecological cliff like the grey partridge. Why are the grey partridges declining so terribly? There is no great mystery. During the Second World War the amount of land under cultivation increased from 12 million acres in 1939 to 18 million in 1945. Towards the end of the forties, the first widely used herbicide went on the market, 2,4-Dichlorophenoxyacetic acid, and herbicides were targeted at arable land. The decrease in wildlife was almost immediate. When the arable weeds went, so did the cereal insects, because the weeds were their host plant. Grey partridge chicks only eat insects; red-legged partridge chicks eat both insects and seeds.

Reared red-legged partridges are also released for shoots at about six million birds a year, some of which escape the guns and go feral, such as the two before me. But because they have been raised in captivity, they recognize a good grain hopper, dispensing food charity, when they see one. The British population is estimated to lie between 90,000 and 250,000 pairs.

Something in the paddock alerts the two red-legs; and they run, bow-backed, into the safety of the wildflower border. Here the corn chamomile is frothing from the earth, and has an unusual orange caterpillar on it. In a nature-watching nadir I try to take a macro photograph of the larva, but the telescoping lens of the camera knocks the hairy caterpillar off the corn chamomile so that it becomes lost in the undergrowth.

7 MAY Swifts arrive, under a sparkling sky. The 'devil's screechers' have timed their arrival to take advantage of the Maytime hatch of insects. Swifts are so sensitive to changing atmosphere that they can sense an area of high pressure miles away. The upwelling of the air as the front approaches draws up insects to the swift's advantage.

A cock pheasant appears out of a slit in the green wheat, luscious and shocking, before disappearing. There are two pheasant hens sitting on eggs in the wood.

I'm not spending as much time as I'd like at Flinders; sheep, cows and chickens elsewhere intervene and preoccupy, so I've lost track of the hares. I'm leaning on the gate, looking over. The dapper chaffinch is perched on the hedge, almost within touching distance, a fat caterpillar for his young squirming in his beak; but he won't go to them with me near. So I go into the field, and as I do so the sparrowhawk shears the air above me, a fledgling of some sort struggling in its right claw. Ludicrously, I want to intervene.

ONE OF THE JILLS has moved into Flinders from the pad-dock, via one of the two 'pop holes' I've made in the fence.

Just as the sun is abdicating, I see her emerge from her new grass cave, sit up, look around, and then nibble at the wheat, which is now deep enough to submerge her when she is belly-flat, ears flat, as hares like to be when eating or moving.

The other jill and the two jacks have remained reso-lutely in the paddock, from which I have removed the last sheep. There is little point in wanting hares then having sheep uncover them for the delectation of every top-of-the-food-chain predator in the 'hood.

Flinders jill has her form a yard or so from the stock fence, in the lanky grass margin. She is not alone in her home-making; the cock red-legged partridge has scraped two shallow nests, thirty yards apart, near the centre of the field. They are cursorily lined with dry grass. The hen red-legged partridge commonly lays two clutches, incubating one herself and leaving the other for the cock. Consequently, each pair has the potential to pro-duce two broods more or less simultaneously. All gamebirds suffer high rates of nesting failure. Their open homes are highly vulnerable to predation from rats, foxes, weasels and birds of prey. Red-legged par-tridges attempt to offset loss by their unusual habit of 'double-clutching'.

The hen has already laid eggs in the first nest, leaving them for the male to incubate. He is taking his time about doing so; the eggs, buff with ash speckles, are exposed to the world. They shine in moonlight.

Edith has been in the cab of the Land Rover for an hour or more, so I take her for a night-time jolly along the lane. The fat hedges have the comfort of a womb wall, and the white saucers of cow parsley float past on the warm air. Bats fly overhead, reaping the insect harvest of the night.

JAUNTY AND JERKY, the summer starling is Flash Harry, as played by George Cole in the St Trinian's films. The bird's coat is more money than taste, with its oily iridescences of purple and green. While the male starling will occasionally manage a musical phrase, between the wheezing and whirring, he is happiest Cockney-whistling.

Although much of the wheat is too deep for them to forage now, the starlings are the pearly kings and queens of the barer patches, where my seeding went astray, and where the early slugs and the rabbits made inroads. The starlings hunt companionably, though they never drop the strut. They are nesting under the roof tiles of Pool Farm, a Georgian farmhouse half a mile up the road and ripe for Kirstie Allsopp. In the meantime, the starlings enjoy its 'unimproved, suitable for restoration' character. The male starlings sit on the gutter singing to deter other males, and snatches of their music sometimes drift down to Flinders on the breeze, as though a pop radio is being tuned in and out.

The swifts nest up at Pool Farm too. Adult swifts cannot take off from the ground, since their scythe-shaped wings, built for speed, produce insufficient lift to raise the bird. Swifts only come down to earth when they are dead. So the swift must nest in high places. Someone will do the farm up one day, and then where will the starlings and swifts go? And I wonder, as I look up at the red-brick farmhouse, whether all planning and redevelopment should come with a clause requiring that buildings provide homes for birds too.

The cross-sky traffic of starlings and swifts, from field to house and back, is so dense and regular that it makes a permanent smoke mark in the daytime sky. Today the cloud has forced the swifts and their *hirondelle* cousins, the swallows, low so they are skimming the wheat and the wildflowers. The first corn poppy is out, and a hundred others have the mouth of their clambuds split open to reveal the shocking scarlet inside. One peculiar detail about poppies: before they flower, they hang their heads exactly like Hector Guimard's Parisian street lamps.

The slender, writhing leaves of cornflowers have something art nouveau about them too, in their curvilinear dynamism; the buds of the plant are tight, jewelled eggs. Quite the Fabergé look.

Looking almost apologetic between the wildflowers I have sown are purple wild pansies.

IN THE WOOD, bluebells quiver between the columns of beech, and the child rooks are making their first faltering caws.

The rook is never a popular subject down at the bar of the Jolly Farmer, unless it is on the menu. 'Branchers' or 'flappers' or 'perchers', so called because they have sufficiently fledged to climb from their nest to the adjoining twigs, are shot for food. In the early twentieth century gun companies such as Holland & Holland and Westley once manufactured special rook rifles, such was the vogue for shooting the birds. Rooks are still shot in parts of the country in the months of April and May.

I have seen it done with a .410 shotgun. In a copse, in east Herefordshire. If the rooks are low it is mere murder; if the rooks are high up in tall trees then one has to sight the dark rook, already diminished by distance, among black flickering leaves; and the wind moves through the tree tops, meaning the bird won't keep still.

SOME TIME IN THE 1970s: The childish .410 pops. By fluke a bird is hit, only, discomfitingly, to aircrash-dive with spread wings to become lodged in the branch below.

I try to climb up but the beech is too fat, too smooth; it might as well be a greased pillar.

None of this happens in silence. From under boots there is the crunch of last autumn's beech-mast shells; the adult rooks are flying in squalling circles. Teenagers shouting. (Children, really.) A friend has borrowed my Paratrooper .22 air rifle and is firing on non-stop automatic into the trees. *Bap-bap-bap-bap-bap. Bap-bap-bap-bap-bap.*

I have a pointless death on my hands, an *acte gratuit*.

We leave with twenty-four black birds, enough to be baked in a nursery-rhyme pie. One dead rook remains behind, stuck for ever, uselessly, in a crook of a beech in Westhide.

I make a promise to myself, as we clamber over the fence, never to waste another bird's life. I will kill for the pot, but it will go in the pot.

There is only one more Johnny-the-Kid blast of vandalism, and it comes a few years later. My friend Jamie and I are coarse fishing on the Wye at Frenchistone, and a fly-fisher with waders keeps walking up and down between our lines, spoiling our sport. Rudely, he asks us to move, although we arrived first.

He's got a suntan, Village People moustache, permed curly hair, which is the same in the 1980s as putting a sign on the head saying, 'I'm in 22 Special Air Service Regiment.' The SAS is based in Hereford, at Bradbury Lines.

Due to school cadets and family associations, we are more Royal Navy.

Jamie says, getting his Webley air pistol out of his fishing bag, 'I bet you can't hit his lure.'

Flip open the breech, push in the soft lead .22 pellet. Safety off. The gun is nicely weighty.

The lure is coming in, spinning, hitting the surface, shining.

Squeeze the trigger. *Phut* of gun, *phlip* of pellet hitting lure.

Jamie actually open-mouthed, me Cool Hand John.

Mr SAS is less impressed, and bellows obscenities. Jamie suggests his language is 'vulgar', at which the moustachioed man runs out of the water, up the bank, picks up his fishing satchel and chases us. We reel in, gather our stuff and light out. For the first thousand yards this is a laugh, since we aren't wearing waders, we are fleet-o'-foot in Adidas trainers.

It turns out that Mr SAS is rather fit, and chases us for a cross-country two miles, and by the time we dart up the track to my house we are pale wraiths of our former selves.

But we lost him half a mile back.

We are lying on the carpet waiting for *Blue Peter* when there is a determined ring on the front-door bell. Some instinct for danger makes me peek out of the sitting-room window. It is the moustachioed man, legs apart on the gravel.

What follows is perhaps the first ever instance of teenagers hiding in front of the sofa.

ROOKS, THE BLACK destroyers of slugs and leather-jackets, are plodding through Flinders. Later, in August, they will pluck at the wheat, a crime for which they were once sentenced to death in Scotland. Overall, I count them the farmer's friend.

ON A DAY WHEN the sun reigns, the first corn marigold blossoms in Flinders; it's a child's version of a flower, round centre with perfectly placed petals around it.

The flower was once such a feature of the British countryside that it supplied the gold in Golding in Shropshire and Goldhanger in Essex. It provided table decorations too. Matthew Arnold wrote to his sister in 1883: 'I thought of you in passing through a cleared cornfield full of marigold. I'll send you one of them, Nelly gathered a handful, and they are very effective in a vase in the drawing room.'

Corn marigold is as old as British agriculture itself, since it was probably brought here by the Neolithic people. Arable farmers, however, have never warmed to its sunny splendour, since the fleshy leaves impeded the harvest reaping. Henry II issued an ordinance against 'a certain plant called Gold', requiring tenants to uproot it, which was probably the earliest enactment demanding the destruction of a weed. In *A Boke of Husbandry*, 1523, John Fitzherbert included 'Gouldes' in his black-list of plants that 'doe moiche harme'.

Herbicides have made Henry II's dream come true. Until today, I have never seen corn marigold growing in a wheatfield.

THE COCK RED-LEGGED partridge never does get to sit on his eggs. They survive a week or more, creamily entic-ing, and then they are taken.

The hen is sitting tight, on a much larger clutch of thirteen eggs. I see them on a late May evening when she goes off for a stroll; it is an evening perfumed by distant elderflower and sung over by blackbirds on two sides of the field, so that when one pauses, the other starts, and the melody goes backwards and forwards, effortless and for ever.

Scientifically, we know evolution exists; spiritually, we know that the nature around us is unimprovable. Can the blackbird's song be bettered?

Or, is there anything more beautiful than a leveret?

THE LEVERETS ARE BORN on 18 May, forty-two days or thereabouts after mating. Flinders jill has two young, born furred and with their brown eyes wide open, and quite able to move.

Only an accident enables me to meet them. Rupert the Border Terrier slips his leash, and in a drama familiar to 'BT' owners I go running after him calling his name, which of course makes him run faster still.

He only stops to snarl at Flinders jill's form; she emerges like a horizontal ballistic missile, and dog and hare race away into the wheat, their passage marked by lines of trembling grass.

Two tiny leverets are in the entrance to the form, still wet with birth fluid or from maternal licking, and are round, and vulnerable in a way that adult hares are not. They are the cute poster animals of British wildlife.

Aesop's fable of 'The Tortoise and the Hare' notwithstanding, hares are the kings and queens of speed. Apart from 'long dogs', lurchers and greyhounds, I have never seen a dog run faster than Rupert, but back he comes, tail-between-his-legs defeated.

We leave the scene.

AFTER BIRTH, MOTHER hares put each leveret in its own form. The naturalist Brian Vesey-Fitzgerald was convinced that the doe carried the leverets to their new forms in her mouth, 'after the manner of a cat carrying kittens'. It is possible that she merely leads them, since they are perfectly able to walk.

Until they are weaned their mother visits them each in turn at night. When the mother is approaching she gives them a low call and their answering calls help her to find them. She then suckles them. The purpose of the separation is obvious when one thinks of the dangers of a nursery on the surface of a field. While she is away, the leverets lie low and still, to avoid detection by predators.

A female can produce three or four litters a year. Flinders jill is boxing with the two jacks from the paddock within a fortnight of giving birth.

I never find out what happens to Paddock jill's first litter, or even if there was one. She does, though, produce a litter in July.

For so when first I in the summer-fields
Saw golden corn
The earth adorn
(This day that sight its pleasure yields),
No rubies could more take mine eye;
Nor pearls of price,
By man's device
Set in enamel'd gold most curiously,
More costly seem to me,
How rich so e'er they be

By men esteem'd; nor could these more be mine
That on my finger shine.

Thomas Traherne, 'The World', c.1660

THE MAY BLOSSOM has fallen, and in the blackthorn of the hedge the sloes which will sustain the thrushes in autumn are already forming.

From the hedge, every couple of minutes, the cock chaffinch darts out, hovers a second with heavy wings, and returns to cover. The chaffinch is fly-fishing, grabbing insects from the tops of the wheat and the corn chamomile which fringes the field in white and yellow.

A blackbird comes full pelt across Flinders to perch on the laneside hedge. The blackbird lacks guile; where it lands, its nest is nearby. In proof, it goes straight down into the hedge, grub in the beak.

Into this admirable scene comes a white cloudburst, puke-and-toffee to smell. Chemical spray from the farm across the lane; and it covers me, it covers the birds.

I'm torn between protesting to the farmer, and the need to get home and showered. I choose the latter.

For once there is no rain. Unlike me, the birds have no shower.

FLAMING JUNE. When I arrive at Flinders, the scarlet pimpernels are closing their pretty heads for the night, so it must be about 3pm. They are not stop-up-late flowers.

You can tell the time by flowers, because some species open and close at particular times. Andrew Marvell, in his poem 'The Garden', 1678, described an elementary 'flower clock':

How well the skilful gardener drew
Of flow'rs and herbs this dial new;
Where from above the milder sun
Does through a fragrant zodiac run;
And, as it works, th' industrious bee
Computes its time as well as we.
How could such sweet and wholesome hours
Be reckoned but with herbs and flow'rs

The Swedish botanist and zoologist Linnaeus devised a full-scale garden plan for a *Horologium Florae* seventy years later, and several botanical gardens implemented it.

British wildflowers are just as reliable as anything planted in a formal garden. In Flinders in June the scarlet pimpernel ('The Shepherd's Clock') opens its petals at 8am, and closes them at 3pm; dandelion opens at 5am and closes at 8pm; and sow thistle opens at 5am and closes at 12pm.

The Reverend James Neil, recounting an English wildflower floral dial in *Rays from the Realms of Nature*, 1879, had lesser celandine opening at 9am, buttercup at 6am.

Here is his complete wildflower 'Floral Dial':

Yelow Goatsbeard,   Opens at III. o'Clock, a.m.
   or Noontide
(*Tragopogon pratensis*)

Wild Succory, or Chicory   "   IV.   "   "
(*Cichorium intybus*)

Common Nipplewort   "   V.   "   "
(*Lapsana communis*)

Buttercup   "   VI.   "   "
(*Ranunculus bulbosus*)

White Water Lily   "   VII.   "   "
(*Nymphaea alba*)

Scarlet Pimpernel   "   VIII.   "   "
(*Anagallis arvensis*)

Proliferous Pink   "   VIII.   "   "
(*Dianthus prolifer*)

Lesser Celandine   "   IX.   "   "
(*Ranunculus ficaria*)

Common Nipplewort   Closes at X.   "   "
(*Lapsana communis*)

Common Star of Bethlehem, or
Lady Eleven O'Clock   Opens at XI.   "   "
(*Ornithogalum umbellatum*)

Yellow Goatsbeard (*Tragopogon pratensis*)	Closes at XII. o'Clock, p.m.		
Proliferous Pink (*Dianthus prolifer*)	"	I.	" "
Scarlet Pimpernel (*Anagallis arvensis*)	"	II.	" "
Rough Dandelion (*Leontodon hispidus*)	"	III.	" "
Wild Succory, or Chicory (*Cichorium intybus*)	"	IV.	" "
White Water Lily (*Nymphaea alba*)	"	V.	" "
Nottingham Catchfly (*Silene nutans*)	Opens at VI.		" "
Evening Primrose (*Oenothera biennis*)	"	VII.	" "

The Reverend Neil was a 'parson naturalist', that curiously British phenomenon which began with the Reverend Gilbert White in Selborne. The great advantage of the Church of England's parochial system was that a parson inclined to nature study could get to know, in minute and comprehensive detail, the plants and animals of one small area of the countryside, often for the whole of his working life. 'Parochial' has become pejorative, when it should have become exalted.

SUNDAY AFTERNOON, just after lunch, and the silence in Flinders is exquisite. The June breeze carefully brushes the green wheat.

In the sky there are white doves, white cloudlets, otherwise the sky is true blue. The cornflower has blossomed, as vividly azure as the ceiling above it. 'Cornflower blue' was long an artist's favourite, the juice from the flowers mixed with alum.

The cornflower once grew throughout the UK and was so common, particularly on sandy, slightly acidic soils, that it became troublesome. Even the eighteenth-century poet John Clare, a friend to almost all nature, wrote of cornflowers 'troubling the cornfields with their destroying beauty'. There was some attempt at weeding by farm workers, who would gather bunches of the vibrant blue flowers and send them up to London flower markets to supplement their meagre wages. They were also used as a buttonhole flower and young men would give these blooms to their sweetheart to divine their chances of success in romance, leading to the country name of 'Bachelor's Button'.

Today the truly native plant is only found on a handful of sites across the country, perhaps the best of all being on the Isle of Wight. What did for the cornflower in its arable habitat, apart from herbicides, was improved techniques in cleaning the seed corn that is to be planted the following year. Once a frequent contaminant of cereals, cornflower seeds were unintentionally drilled each year along with the crop, thereby ensuring that they were spread around the local area. With twentieth-century improvements in seed-cleaning technology, cornflower seeds were removed before the seed corn was planted, ending the inadvertent sowing system so crucial to the little blue flower's survival.

The French still have the cornflower in cornfields; the *bleuet* is their flower of war remembrance, the equivalent of our Flanders poppy. The cornflower is the national flower of Germany and the reason it was chosen is rather lovely. When Napoleon forced Queen Louise of Prussia from Berlin, she hid her children in a cornfield and kept them entertained and quiet by weaving wreaths of cornflowers. One of her children, Wilhelm, later became the Emperor of Germany. Remembering his mother's bravery, he made the cornflower a national emblem of unity.

4 June By moving my straw-bale hide over to the side of the field where Flinders jill has secreted her leverets, I, on this enchanted evening, see them suckling.

I had supposed she would feed them individually, but they scamper out to meet her and suckle simultaneously, hanging from her front. Swifts skirl and banshee.

5 JUNE The kestrel sits lazily on the telephone wires crossing Flinders, an aerial station which saves hovering.

I know what he is watching in the wheat below. An adult red-legged partridge is *ku-ku*king a frantic alarm. The kestrel launches off, hovers six feet above the ground, then pounces. He flies away to feed his young with a red-legged partridge chick. A bird for a bird.

The chicks hatched yesterday, maybe the day before. I wanted scenes like this, kestrels hunting over flower-embroidered, running-with-hares corn. And I have got them.

Still, somehow it hurts to lose one of the partridge chicks. Nature is running red in beak and claw.

There are eight partridge chicks left. Later I hear them cheeping in the wheat, close by.

Hoverflies hang in unmoving air.

10 JUNE I'm not convinced that Flinders jill always 'mews' to her leverets to alert them to suckling time; today, at about 8pm, she silently lumbers past their forms, then waits fifteen yards to the front. I think the leverets smell and sight her. Today while they feed, she is wired stiff with all-seeing, all-hearing concentration on danger. Her long bristles catch the breeze, and they are her only moving parts.

Hares are capable of superfoetation – that is, the fertilization of new eggs while already pregnant, giving a potential minimum duration between successive litters of 38 days. One of the males has been hanging around Flinders jill, so she may already be pregnant with a new litter.

Before and after suckling, the leverets have taken to playing, jumping in the air, running in circles and tilting at an unfortunate red-legged hen.

During the spraying sixties, hare numbers crashed by a million or more. There are perhaps 730,000 brown hares left in the UK. Well, 730,002 now. Like many other fauna in these isles, the hare is running for its life.

WATCHING WHEAT GROW, close up, has an aesthetic satisfaction. First a green and tender blade, the wheat sprawls, then shoots up straight, and the ear starts to form. By and by, when every stalk is tipped, in Richard Jefferies' memorable image, 'like a sceptre', the stalk is still green but the ear is fading into yellow.

Today, on 20 June, the wheat has reached this sceptre stage. There are chaffinches and starlings (the young ones grey compared to their elders) in the wheat searching for bugs, and six red-leg partridge chicks are running hidden as a human would be in a forest. A pair of collared doves, delicate and betrothed, have moved into the wood and are using Flinders for dining. The lovebirds call to each other constantly; three phrases played on a recorder with a broken reed: *coo-coo-coohek.*

There is a whole world of life in my cornfield. There are field mice and voles, nameless and numberless flies, and bees galore. Meadow brown butterflies are on the wing; the chocolate-brown males of the species are more common, and try to drive rivals away from the females sitting on purple thistles in the corner. There is, it has to be admitted, no better plant for nature than the despised, sow-itself-everywhere thistle. Already the plant's seed heads, which promise food for autumn finches, are forming.

The thistle is only one of a catalogue of wildflowers – 'weeds', if you will – that have sown themselves in Flinders. So far I have recorded:

Scarlet pimpernel
Forget-me-not
Speedwell
Thistle
Hairy bittercress
Wild pansy
Hedge bindweed
Shepherd's purse
Wintercress
Groundsel
Sun spurge
Docks
Good King Henry
Wild oats

The rampant groundsel is already white and hairy with seed; the heads are the beards of senile men, hence its generic Latin name, *Senecio*, from *senex*, or old man. I lean over into the wheat and pull out handfuls to take home. Groundsel was once grown as a crop for pigs and poultry. I have both.

The seeds of these listed flowers have either survived modern farming, blown in on the wind, or arrived via the faeces of birds. According to Jack Evans, the retired farm labourer who lives at Brook Cottage along the lane, Flinders was ploughed up (from permanent pasture) in about 2000. ('I don't know why, with clay like that; been better kept for stock.') He brings along a black-and-white aerial picture of Brook Cottage taken in 1967, when some enterprising pilot flew all over the Herefordshire countryside snapping properties from above, and selling the photos, in frames. (We've got one of our house.) In the bottom-left corner of the photograph is Flinders, with a tractor doing a hay cut.

Jack Evans is a regular at Flinders gate, and has developed a soft spot for Rupert the Border Terrier, whom he calls 'the little lion'.

I HAVE A SORT of control for my Olde Wheatfielde/ future-of-farming experiment: the Chemical Brothers' wheatfield next door, which is uniform in colour, except for some clouds of cow parsley that have proved

152

resistant to repeated chemical-dousing. There are no butterflies fluttering, no songbirds singing, only tomb-time stillness and silence. The minuscule field margin, which might have provided a habitat for something, has been scythed down to lawn-like height in preparation for combining.

One day I watch a female pheasant emerge from the wood with her five stripey chicks; she leads them down the side of the Chemical Brothers' wheatfield; she tries to lead her young into the wheat but it is so dense it is barred to her.

The 2007 Countryside Survey found that, during the previous decade alone, the species richness of British fields had declined by 8 per cent.

Compare the deadness of the Chemical Brothers' wheatfield, which can stand representative for 90 per cent of modern cereal land, with what Richard Jefferies wrote about a Victorian cornfield:

Let your hand touch the ears lightly as you walk – drawn through them as if over the side of a boat in water – feeling the golden heads. The sparrows fly out every now and then ahead; some of the birds like their corn as it hardens, and some while it is soft and full of milky sap. There are hares within, and many a brood of partridge chicks that cannot yet use their wings. Thick as the seed itself the feathered creatures have been among the wheat since it was sown. Finches more numerous than the berries on the hedges; sparrows like the finches multiplied by finches [sic], linnets, rooks, like leaves on the trees, wood-pigeons whose crops are like bushel baskets for capacity; and now as it ripens the multitude will be multiplied by legions, and as it comes to the harvest there is a fresh crop of sparrows from the nests in the barns, you may see a brown cloud of them a hundred yards long . . . There are, then, the poppies, whose wild brilliance in July days is not surpassed by any hue of Spain. Wild charlock – a clear yellow – pink pimpernels, pink-streaked convolvulus, great white convolvulus, double-yellow toadflax, blue borage, broad rays of blue chicory, tall corn-cockles, azure corn-flowers, the great mallow, almost a bush, purple knapweed – I will make no further catalogue, but there are pages more of flowers, great and small, that grow at the edge of the plough, from the coltsfoot that starts out of the clumsy clod in spring to the white clematis.

If I had a plan, a guide, a picture of what Flinders should be, it is Jefferies' Victorian walk in a wheatfield.

MIDDAY AT FLINDERS: I sit in the Land Rover cab while it rains. The chiffchaffs and the blackbirds are dismal and quiet. Only the chaffinch babies outdo the rain with their gaping clamour.

Earlier, meadow brown and small tortoiseshell butterflies had been flying, but in rain butterflies die.

There are rats in the wheat; they have their burrows in the hedge. In a piece of environmentally friendly rat control, nature hitting back as it were, the fat rodents, waddling slow, are pounced on by the vixen from Three Acre Wood. She leaves their tails behind as proof; the discarded appendages look like small snakes. Still the rats raid the wheat. There is one in there now. Bedraggled; the proverbial drowned rat.

I like the fox to take rats. I do not like her to take hares. As if she minds.

Corn needs rain. The roots of a single stalk, it is said, make a quarter mile if put end to end. A single acre of corn will lift 250 tons of water between sowing and harvesting. Alas, corn also develops mildew with an abundance of moisture. Some of my corn, in the wettest, lowest corner, is affected by mildew as black as the clouds overhead.

Down in the same corner, the vixen slides through the hedge and under the wire to stalk the wheat, as a tiger stalks the jungle.

The vixen is monomania in canine guise: I keep reinforcing the fencing against her, she keeps digging under.

There is cat and mouse, there is farmer and fox. We are playing out an endless game, an eternal rivalry.

In its youthful development, the wild oat cannot be distinguished from cultivated wheat. It is not until now, as the harvest begins to ripen, that I can see the intruder. The wild oats' grains do not point upwards in clusters but hang gracefully downwards from the slenderest horizontal arms. There is a terrible pagan beauty to wild oats, siren and kinetic above the wheat. In a certain silhouetting light, though, they are delicate Chinese calligraphic strokes on the dusk.

There are fifty or so wild oat plants in the field; I hand rogue those around the edges. The wild oat is the curse of crop farmers. One reason for the plant's invasive success is that it can sow itself: it ripens and sheds its grain before the corn is harvested; on the oat grain is a long hygroscopic awn, kinked in the middle. This awn (a short, dark and uncommonly animalish bristle) twists, straightens, moves in reaction to changes in humidity. Eventually, the grain insinuates itself into a crevice in the soil.

That is the science; to watch the wild oat perform its self-sowing strategy, put a wild oat grain on the cuff of a woollen jumper. Within half an hour, the main awn, together with the small bristles at the base, will have 'walked' the grain around and up the sleeve almost to the elbow.

Farmers were told to diversify, and for years I've drawn up elaborate schemes for farmhouse B&B and self-catering. I've even got as far as planning yurts. Penny has never been keen.

Eventually she coughs it up. 'You've got as many hospitality skills as Basil Fawlty.'

Perhaps, then, I suggest, I should get a part-time office job?

You read about people corpsing, but I've never actually seen it before. Penny bends double with laughter, actually clutching her stomach with the pain of it.

She starts to tell me what is so funny, but then breaks down again, this time with rolling tears of jollity.

When she can finally breathe and speak she explains the joke. Apparently, the idea of me doing office politics is the most amusing thing in the entire history of amusing things. My patience and charm, such as they are, have a span of about three hours. The lifetime of a fruit fly.

What jobs require the social skills of Simeon Stylites, he who sat on top of a pillar for thirty years? Alone.

Oh, I know. Farming.

Oh, I know. Writing. Consequently, you are reading this book. Sometimes I put on a tin hat and write about military history, and this is how I come to spend a June week in London researching in the archives of the Imperial War Museum, Lambeth. (Perhaps I have the last laugh? I'm in the capital, which my wife yearns to visit, being a London girl, and she is left in the sticks feeding sheep.)

I stay in a 'Superior Loft Apartment' in Barnes. Of course it is booked via Airbnb, the latest middle-class fad. Barnes? Because everyone tells me it is 'leafy' and I'm well known for liking green spaces.

By Day 3 I am going quietly mad(der).

Dear London, do you have any idea of what you are like? At 6am the jets for Heathrow start coming over on an unerring flight path across the bottom-left corner of the Velux above my bed.

White jet. Wait a minute. White jet. Wait a minute. White jet . . . I lie there waiting for a bird to fly over. None does. In the Middle Ages, swifts clamoured around London buildings, and sparrows were thick on the ground. So, is this progress to have desert for sky? I live by the river (yes, I know my Clash lyrics) for a week and wonder: when did the right for a human's cheap holiday trump the right to quietude, and a bird's right to fly?

On Day 4 I'm down in Lambeth North tube station with the rest of the IWM archive-trawlers, just after the research rooms have closed. A black guy with a red-gold-green headband comes on to the platform, and is somewhat befuddled seeing the incongruous white people. He throws his arms out, and shouts, 'I want to see some black people here!'

I feel for him. When I get out at Barnes, I feel like shouting, 'I want to see some birds here!'

Day 5. I take the train down to Surbiton to see my godparents, Edward and Pru. We sip thin ginger tea from bone-china cups, and eat fat slices of fruit cake.

From the sitting room I have a view down the garden, with its well-trimmed lawn, its pretty flower border . . . it is the exemplar of Metroland gardens. John Betjeman would smile approvingly. Except for this: in two hours or so the bird tally is one wood pigeon. Edward must have read my thoughts, for he said suddenly, in his bassoon Indian Army voice, 'No birds now, of course. The cats have got them all. When we moved here forty years ago there were *loads* of birds.'

The domestic and feral cat population of GB has doubled in the period. Of course, the garden of neighbouring Number 88 hardly helps the Surbiton wildlife. In Number 88 Betjeman would surely see a little Slough, and have cause to verse:

> *Come friendly bombs and fall on the decking next door!*
> *It's such an eyesore*
> *There isn't grass for a probing jackdaw*
> *Swarm over, Death!*

Day 6, I sit on the terrace of the White Hart pub drinking an evening pint of Fuller's London Pride. A woman – and I have to do a double take at this – wanders along the towpath with a pet miniature black-and-white pig. His name is Walter, and he gets fed crisps by everyone; it's a walk-through dining service.

Walter is the highlight of my week in London. Still, as they keep saying, you can take the boy off the farm, but you can't take the farm out of the boy.

On the Intercity 125 train home, going through lovely Oxfordshire, it occurs to me that George Monbiot's re-wilding is an idea which should, in the argot, 'get in the fucking sea' with the red herrings. Lynx. Beaver. Wolf. What is the connection between these oft-touted possible reintroductions to the British landscape? They are all charismatic species, and only suitable for the remotest corners. Re-wilding is at best fiddling at the edges of Britain's environmental problems, at worst an absolute diversion from them. The total area of the UK under agricultural production is 18.7 million acres, or 71 per cent of the overall land surface, of which roughly a third (15.5 million acres) is arable, the rest being various types of grassland. On Britain's farmland something like an ecological holocaust is taking place.

Don't take my word for it. Take that of the Department for Environment, Food and Rural Affairs, publisher of *Wild Bird Populations in the UK, 1970 to 2014.* Regarding farmland birds:

- In 2014, the breeding farmland bird index in the UK was less than half (a decline of 54 per cent) of its 1970 level – the second lowest level recorded.
- Within the index over the long term period, 21 per cent of species showed a weak increase, 21 per cent showed no change and 58 per cent showed either a weak or a strong decline.
- Most of the decline for the farmland bird index

occurred between the late seventies and the early nineties, largely due to the impact of rapid changes in farmland management during this period.

- The smoothed indicator shows a significant on-going decline of 11 per cent between 2008 and 2013.

The farmland bird index comprises 19 species of bird. The long term decline of farmland birds in the UK has been driven mainly by the decline of those species that are restricted to, or highly dependent on, farmland habitats (the 'specialists'). Between 1970 and 2014, populations of farmland specialists declined by 69 per cent while farmland generalist populations declined by 9 per cent. Changes in farming practices, such as the loss of mixed farming systems, the move from spring to autumn sowing of arable crops, and increased pesticide use, have been demonstrated to have had adverse impacts on farmland birds such as skylark and grey partridge, although other species such as wood pigeon have benefited. Four farmland specialists (grey partridge, turtle dove, tree sparrow and corn bunting) have declined by 90 per cent or more relative to 1970 levels. By contrast two farmland specialists (stock dove and goldfinch) have more than doubled over the same period, illustrating how pressures and responses to pressures vary between species. Overall, 75 per cent of the 12 specialist species in the farmland indicator have declined over this long term period, while 17 per cent have increased and 8 per cent have shown no change.

LONG TERM CHANGE (1970–2013)

Species	Long term % change	Annual % change	Trend
corn bunting (*Emberiza calandra*)	–91	–5.34	strong decline
goldfinch (*Carduelis carduelis*)	146	2.11	weak increase
grey partridge (*Perdix perdix*)	–92	–5.62	strong decline
lapwing (*Vanellus vanellus*)	–66	–2.46	weak decline
linnet (*Carduelis cannabina*)	–60	–2.1	weak decline
skylark (*Alauda arvensis*)	–60	–2.13	weak decline
starling (*Sturnus vulgaris*)	–81	–3.78	strong decline
stock dove (*Columba oenas*)	102	1.65	weak increase
tree sparrow (*Passer montanus*)	–90	–5.21	strong decline
turtle dove (*Streptopelia turtur*)	–97	–7.67	strong decline
whitethroat (*Sylvia communis*)	1	0.01	no change
yellowhammer (*Emberiza citrinella*)	–55	–1.83	weak decline

SHORT TERM CHANGE (2008–2013)

Short term % change	Annual % change	Trend
-12	-2.52	weak decline
30	5.33	strong increase
-23	-5.17	strong decline
-37	-8.73	strong decline
-6	-1.25	weak decline
-14	-3.08	strong decline
-19	-4.1	strong decline
18	3.32	strong increase
34	5.96	strong increase
-68	-20.51	strong decline
17	3.16	strong increase
0	004	no change

The response of the main farmers' organization, the National Farmers' Union? @NFUtweets trilled: 'British farmers help farmland birds in so many different ways. Please RT & share our poster #BackBritishFarming.'

The poster was headed: 'How Farmers Help Farmland Birds'. According to the poster, farmers '*leave stubble in the field over winter to encourage growth of weeds that support insects*'.

That clapping sound? The ghost of Goebbels putting his hands together in awed appreciation of some real propaganda. About 3 per cent of arable land, as already mentioned, is given over to winter stubble.

The principal EU subsidy to British farmers, the Basic Farm Payment, requires them to do nothing more than observe the environmental protection legislation they are anyway legally bound to observe.

Look over the specialist farmland list again. One of the quintessential birds of farmland, the corncrake, is now so rare that Defra does not bother to list it. Ditto the quail.

*

DOWN TO FLINDERS I go on my first evening back from London, the lanes almost blocked by cow parsley. When the Land Rover bashes it, the bitter green scent comes in through the open windows.

When I get to the gate at Flinders I can scarcely believe my eyes. Or contain my wonder. The wildflowers are out in force, with poppies, corn chamomile, cornflower and corn marigold lighting up the wheat. The 'borders' look like the page of an illuminated manuscript, or a stained-glass window in a cathedral.

I can scarcely believe my ears. This kingdom of flowers buzzes with flies, crickets and grasshoppers, and hums with bees. (Though I notice that some bees positively warble, oddly reminiscent of Charles Aznavour singing 'She'.) It is *loud*.

Wheat wobbles with shield bugs, blackbirds are pogoing under the hedge, there are squadrons of starfighter starlings in and out of the field, the hen pheasant and her two remaining chicks have moved in. Across the green, growing expanse of wheat I can hear a partridge calling.

The field is alive. In motion. Vibrating with life.

But the hares? I am sufficiently concerned about my hares to wait till dusk, a long wait on a day just past midsummer. For once I switch on the torch, and there they are: Flinders jill and her two leverets, the latter now half grown. They have lost their infant cuteness, and have assumed the angular haughtiness of the adult. Nonetheless, they are running in playtime circles.

The leverets no longer suckle, and are technically independent of their mother. The family is sticking together because there is always safety in numbers.

RAINING JULY: I don't need this much rain. Or this much wind. The starlings are whipped off the ground and away to somewhere more sheltered.

Charles II wittily described the English summer as 'three fine days and a thunder storm'. The ditch is full, of water plus the soup of pollutants from the Chemical Brothers' wheatfield. From Flinders, the ditch water makes its way to the Garron, then to the Wye, then to the Severn, then to the Irish Sea, then to the Atlantic, then to all the seas of the world . . .

Aside from telling the time, the scarlet pimpernel is a weather forecaster, as the poor mad peasant poet John Clare noted in his *Shepherd's Calendar*:

> *And scarlet starry points of flowers*
> *Pimpernel dreading nights and showers*
> *Oft calld 'the shepherds weather glass'*
> *That sleep till suns have dryd the grass*
> *Then wakes and spreads its creeping bloom*
> *Till clouds or threatning shadows come*
> *Then close it shuts to sleep again*

In a sense, I had John Clare at my wedding, in that tiny church, a ship in a galing sea, at Llanthony in the Black Mountains. 'First Love' it was, read by a second cousin:

I ne'er was struck before that hour
    With love so sudden and so sweet,
Her face it bloomed like a sweet flower
    And stole my heart away complete.
My face turned pale as deadly pale,
    My legs refused to walk away,
And when she looked, what could I ail?
    My life and all seemed turned to clay . . .

By then Clare had been dead for 150 years, although more relevant than ever, and not purely because he could delineate *amour*. John Clare was the first environmentalist poet.

In poems such as 'The Lament of Swordy Well' and 'The Mores' Clare, the son of a farm labourer and no fey middle-class Romantic, documented the effect of enclosure, granted to local landowners, on what had been open heath and moor available to all. Before the Acts of Enclosure:

Unbounded freedom ruled the wandering scene
Nor fence of ownership crept in between
To hide the prospect of the following eye
Its only bondage was the circling sky
One mighty flat undwarfed by bush and tree
Spread its faint shadow of immensity
And lost itself, which seemed to eke its bounds
In the blue mist the horizon's edge surrounds
Now this sweet vision of my boyish hours

Free as spring clouds and wild as summer flowers
Is faded all – a hope that blossomed free,
And hath been once, no more shall ever be
Inclosure came and trampled on the grave
Of labour's rights and left the poor a slave
And memory's pride ere want to wealth did bow
Is both the shadow and the substance now . . .

The ancient elms were felled, marshes drained, rivers canalized, and the fields squared and regimented. Farming became more profitable with enclosure, except for the poorer sort of farmer and the labourers, especially those who depended on the commons for their survival. They were deprived of their living.

Yet it was not just humans who suffered under enclosure. So did nature. Intensive usage of the land was to the detriment of the snipe who dwelled in the mire, the glow worms who lit the heath, the corncrake who hid among the crops:

And birds and trees and flowers without a name
All sighed when lawless law's enclosure came . . .

All that Clare loved was torn away. With the destruction of his heartland went the deterioration of his own mind. Bouts of depression on top of poverty, dismay at enclosure, seven children to feed, a forced move from his childhood village of Helpston, felled him. He moved only three miles to Northborough, yet it was enough: 'I have had some difficulties to leave the woods & heaths & favourite spots that have known me so long for the very molehills on the heath & the old trees in the hedges seem bidding me farewell.' Clare and his landscape were one; take away one, lose the other. (His poetry thereafter became preoccupied with nests.) At the age of thirty, he was put into an asylum near London. Four years later he walked out one July morning and in split pauper's shoes made his way back home, eighty miles, sleeping in barns, his head towards the north to show himself the steering point in the morning, living on a chew of tobacco, grass, and a sup of ale when some kind stranger tossed him a penny. After a few months, 'homeless at home', he was confined in Northampton General Lunatic Asylum for his last twenty-four years.

I PLAY AN INTERNET game that is simultaneously amusing and alarming: Google any arable or 'tillage' flower, such as scarlet pimpernel or poppy, and the results will feature the said flower designated as a 'weed'. You will also get advice from an agrochemical company on how to kill it; after all, as Bayer Crop Science UK warns: 'Weeds are a constant threat. They can interfere with crop growth and limit yield potential. Bayer's herbicides fight weeds with a vengeance; controlling weed pressure and providing reliable, season-long control and burndown solutions. These herbicides may utilise multiple modes of action to help combat glyphosate-tolerant and resistant grass and broad-leaf weeds.'

Fortunately, Bayer has an arsenal of chemical weapons for weed-cleansing, many of them branded downright charmingly. Among Bayer's herbicides are Artist, Atlantic, Harvest, Liberator, Pacifica, Regatta. My favourite, though, is Othello: 'Othello is the product we recommend for outstanding post-emergence control of annual meadow grass and broad leaved weeds in winter wheat. It is the first oil dispersion herbicide that we have launched in the UK.' Oh, the irony. Othello the herbicide would kill the flora mentioned in *Othello* the play. For Shakespeare, the corn poppy is the balm that brings 'sweet sleep'; in Bayer's version of Othello, the corn poppy is a menace to the harvest.

The Bard was quite the fan of flowers, and his plays display an extensive knowledge of plants, their purposes and the superstitions attached to them, as Henry Ellacombe detailed as long ago as the 1880s with his *Plant-Lore & Garden-Craft of Shakespeare*.

On an idle evening I leaf through *Plant-Lore* in a tallying of Shakespeare's flowers which would be killed off by Othello the herbicide:

CHAMOMILE – HENRY IV, PART 1

> *Falstaff: 'For though the camomile, the more it is trodden on the faster it grows, yet youth, the more it is wasted the sooner it wears.'*

CORNCOCKLE – CORIOLANUS

Act 3, Scene 1: *Coriolanus and the senators of Rome argue about a free gift of corn to the people; the gift is likened to the nourishing of the 'cockle . . . which we ourselves have plough'd for, sow'd, and scattered', and which Coriolanus claims will incite 'rebellion, insolence, sedition' among the common people.*

CORNFLOWER (BACHELOR'S BUTTON) – Merry Wives of Windsor

Hostess: *'What say you to young Master Fenton? He capers, he dances, he has eyes of youth, he writes verses, he speaks holiday, he smells April and May; he will carry 't, he will carry 't; 'tis in his buttons; he will carry 't.'*

CROW-FLOWER (RAGGED ROBIN) – HAMLET

Gertrude:
*'There with fantastic garlands did she come
Of crow-flowers, nettles, daisies, and long purples.'*

CUCKOO-BUDS (CELANDINE) – LOVE'S LABOUR'S LOST

*'When daisies pied and violets blue
And lady-smocks all silver-white
And cuckoo-buds of yellow hue
Do paint the meadows with delight.'*

CUCKOO-FLOWER/LADY'S SMOCK (CARDAMINE PRATENSIS) – KING LEAR

Cordelia:

171

'He was met even now
As mad as the vex'd sea – singing aloud;
Crown'd with rank fumiter and furrow weeds,
With burdocks, hemlock, nettles, cuckoo-flowers,
Darnel, and all the idle weeds that grow
In our sustaining corn.'

WILD DAFFODILS – A WINTER'S TALE
Perdita:
'Daffodils
That come before the swallow dares, and take
The winds of March with beauty.'

DAISIES – LOVE'S LABOUR'S LOST

DARNEL (CORNCOCKLE) – KING LEAR

FERN – HENRY IV, PART 1
Gadshill: 'We have the receipt of fern-seed, we walk invisible.'
Chamberlain: 'Nay, by my faith, I think you are more beholding
to the night than to fern-seed for your walking invisible.'

HAREBELL – CYMBELINE
Arviragus:
'Thou shalt not lack
The flower that's like thy face, pale primrose, nor
The azured harebell, like thy veins.'

HOLY THISTLE – MUCH ADO ABOUT NOTHING
Margaret: 'Get you some of this distilled Carduus Benedictus,

and lay it to your heart; it is the only thing for a qualm.'
Hero: 'There thou prickest her with a thistle.'
Beatrice: 'Benedictus! why Benedictus? You have some moral in this Benedictus.'
Margaret: 'Moral! no, by my troth, I have no moral meaning: I meant plain holy-thistle.'

KNOTGRASS – A MIDSUMMER NIGHT'S DREAM

Lysander:
'Get you gone, you dwarf,
You minimus, of hindering knot-grass made,
You bead, you acorn!'

LONG PURPLE (FOXGLOVE) – HAMLET

Part of Ophelia's garland

LOVE-IN-IDLENESS (WILD PANSY) – A MIDSUMMER NIGHT'S DREAM

Oberon:
'Yet mark'd I where the bolt of Cupid fell:
It fell upon a little western flower,
Before milk-white, now purple with love's wound,
And maidens call it love-in-idleness.'

MUSTARD – HENRY IV, PART 2

Falstaff: 'wit as thick as Tewkesbury mustard'

NETTLES – KING LEAR

PIGNUT – *THE TEMPEST*

Caliban:

'I prithee, let me bring thee where crabs grow;
And I with my long nails will dig thee pignuts.'

PLANTAIN – *ROMEO AND JULIET*

Benvolio:

'Take thou some new infection to thy eye,
 And the rank poison of the old will die.'
Romeo: 'Your plantain leaf is excellent for that.'
Benvolio: 'For what, I pray thee?'
Romeo: 'For your broken skin.'

Also *TWO NOBLE KINSMEN*

Palamon:

'These poor slight sores
 Need not a plantain.'

PRIMROSE – *HAMLET*

Ophelia:

'Like a puffed and reckless libertine,
 Himself the primrose path of dalliance treads,
 And recks not his own rede.'

Also *MACBETH*

The porter: 'I had thought to have let in some of all
professions that go the primrose way to the everlasting
bonfire.'

Spear-grass (couch grass) – Henry IV, part 1
   Used for tickling the nose to make it bleed

Violets – Hamlet
   Laertes:
   'Lay her i' the earth:
   And from her fair and unpolluted flesh
   May violets spring.'

Also **Pericles**
   Marina:
   'The yellows, blues,
   The purple violets and marigolds,
   Shall as a carpet hang upon thy grave,
   While summer-days do last.'

And **The Winter's Tale**
   Perdita:
   'Violets dim,
   But sweeter than the lids of Juno's eyes
   Or Cytherea's breath.'

Is it just me or would Shakespeare's plays have been poorer without the flowers and plants?

It is still raining, and I am still complaining. Too much water before harvest is a disaster, as Thomas Hardy explained in novel form in *The Mayor of*

Casterbridge. The ears of wheat sprout before they are cut.

Mind you, if it was too dry I'd tell you about the legendary long hot summer of 1976, when grain was as hard and thin as gramophone needles.

You have to know that farmers are constitutional complainers, as rural writer (and farm boy) Richard Jefferies pointed out a century and a half ago:

> 'We be all jolly vellers what vollers th' plough!' – but not to listen to, and take literally according to the letter of the discourse. It runs something like this the seasons through as the weather changes: 'Terrible dry weather this here to be sure; we got so much work to do uz can't get drough it. The fly be swarming in the turmots – the smut be on the wheat – the wuts be amazing weak in the straw. Got a fine crop of wheat this year, and prices be low, so uz had better drow it to th' pigs. Last year uz had no wheat fit to speak on, and prices was high. Drot this here wet weather! the osses be all in the stable eating their heads off, and the chaps be all idling about and can't do no work: a pretty penny for wages and not a job done. Them summer ricks be all rotten at bottom . . .'
> And so on for a thousand and one grumbles, fitting into every possible condition of things, which must not, however, be taken too seriously; for of all other men the farmer is the most deeply attached to the labour by which he lives, and loves the earth on which he walks like a true autochthon. He will not leave it unless he is suffering severely.

*

SOMETIMES WHEN I go to Flinders I almost burst out laughing. A wave of a Disney princess's wand would fail to make a scene more colourful or alive.

Bees land and rise in mass Mexican waves on the flower border, which trembles from below, too, with the cryptic movement of toads. If I stand on the far side of the field, the insect buzz and birdsong is such a strong haze it negates the traffic noise on the lane.

Where did the toads come from? There is not a pond or a ditch to breed in within half a mile. Medieval peasants believed that some flowers, such as corn chamomile, were not spread by seed, but were, rather, part of the earth, like rocks. Perhaps the toads were always there, dormant in the soil.

Poor toads, the beasts in the beautiful flowers, the latter predominantly red, white and blue, lending a distinctly patriotic, Union Jack effect to the scene. Before herbicides, the fields of England, meadow and arable alike, vibrated with summer colour. And now? I look around and, except for the exotic Island of Flinders, the countryside is as monotone in summer as it is in winter.

IN THE LAST FIFTY years, ten wildflowers have been 'lost', as the euphemism has it, from the British countryside.

RIP narrow-leaved cudweed; summer ladies' tresses;

small bur-parsley; purple spurge; lamb succory; interrupted brome; downy hemp-nettle; Irish saxifrage; stinking hawksbeard; York groundsel.

Arable plants are the most critically threatened group of wild plants in the UK. Flowers familiar to farmers since the Stone Age have been brought to the edge of extinction, and over into Nothingness.

Farming practices were developed to eliminate 'weeds', and they have done it with Final Solution efficiency. Victorian technology that allowed better seed-cleaning caused the first decline in arable plants; but it was herbicide development in the 1940s that brought holocaust to arable land. For bad additional measure, increases in nitrogen application and the development of highly competitive crop varieties put additional pressure on numerous arable plants (the latter probably did for lamb succory). More, the sieves on modern combine harvesters became nearly 100 per cent efficient in removing weed seeds. (The complaint that my wildflowers will 'contaminate' crops is baseless: combines and seed-cleaners mean any harvest can easily be made acceptable to bakers and seed merchants.)

Some arable flowers found a refuge under the hedgerow. But then the hedges were grubbed out to facilitate the use of large machinery. Only half the length of hedgerow present in Britain in 1945 was still there in 1990.

Politics played its role as well. The effect of the

1947 Agriculture Act and Britain's adoption of the Common Agricultural Policy was to encourage the intensification of arable farming by subsidizing production while guaranteeing markets for surpluses. It was up, up, up with agribusiness, agrochemicals, artificial fertilizers, and farm mechanization. And down, down, down with ploughland's fauna and flora.

THERE ARE – OR RATHER WERE – 150 British flora characteristic of the arable environment. The list of the extinct and the threatened is:

Extinct
Lamb succory (*Arnoseris minima*)
Interrupted brome (*Bromus interruptus*)
Thorowax (*Bupleurum rotundifolium*)
Small bur-parsley (*Caucalis platycarpos*)
Downy hemp-nettle (*Galeopsis segetum*)

Extinct in arable habitats
Narrow-leaved cudweed (*Filago gallica*)
Darnel (*Lolium temulentum*)

Critically Endangered
Upright goosefoot (*Chenopodium urbicum*)
Red hemp nettle (*Galeopsis angustifolia*)
Corn cleavers (*Galium tricornutum*)

Corn buttercup (*Ranunculus arvensis*)
Shepherd's-needle (*Scandix pecten-veneris*)

Endangered
Pheasant's-eye (*Adonis annua*)
Ground-pine (*Ajuga chamaepitys*)
Corn chamomile (*Anthemis arvensis*)
Red-tipped cudweed (*Filago lutescens*)
Broad-leaved cudweed (*Filago pyramidata*)
Corn gromwell (*Lithospermum arvense*)
Grass-poly (*Lythrum hyssopifolium*)
Annual knawel (*Scleranthus annuus*)
Small-flowered catchfly (*Silene gallica*)
Spreading hedge-parsley (*Torilis arvensis*)
Narrow-fruited cornsalad (*Valerianella dentata*)
Broad-fruited cornsalad (*Valerianella rimosa*)
Fingered speedwell (*Veronica triphyllos*)
Spring speedwell (*Veronica verna*)

Vulnerable
Stinking chamomile (*Anthemis cotula*)
Rye brome (*Bromus secalinus*)
Nettle-leaved goosefoot (*Chenopodium murale*)
Corn marigold (*Chrysanthemum segetum*)
Common ramping-fumitory (*Fumaria muralis* ssp. *neglecta*)
Fine-leaved fumitory (*Fumaria parviflora*)

Few-flowered fumitory (*Fumaria vaillantii*)
Large-flowered hemp-nettle (*Galeopsis speciosa*)
Henbane (*Hyoscyamus niger*)
Smooth cat's-ear (*Hypochaeris glabra*)
Wild candytuft (*Iberis amara*)
Yellow vetchling (*Lathyrus aphaca*)
Weasel's-snout (*Misopates orontium*)
Mousetail (*Myosurus minimus*)
Cat-mint (*Nepeta cataria*)
Prickly poppy (*Papaver argemone*)
Night-flowered catchfly (*Silene noctiflora*)
Corn spurrey (*Spergula arvensis*)
Perfoliate pennycress (*Thlaspi perfoliatum*)
Slender tare (*Vicia parviflora*)

Additional rare arable species
Corncockle (*Agrostemma githago*)
Hairy mallow (*Althaea hirsuta*)
Small Alison (*Alyssum alyssoides*)
Annual vernal-grass (*Anthoxanthum aristatum*)
Cornflower (*Centaurea cyanus*)
Purple viper's bugloss (*Echium plantagineum*)
Western ramping-fumitory (*Fumaria occidentalis*)
Purple ramping-fumitory (*Fumaria purpurea*)
Martin's ramping-fumitory (*Fumaria reuteri*)
False cleavers (*Galium spurium*)
Smaller tree-mallow (*Lavatera cretica*)

Field cow-wheat (*Melampyrum arvense*)
Greater yellow-rattle (*Rhinanthus angustifolius*)
Cut-leaved germander (*Teucrium botrys*)

Source: C. M. Cheffings and L. Farrell, *The Vascular Plant Red Data List for Great Britain*, 2005/UK Biodiversity Action Plan

ON THE LANE. We're driving home, and there's a crow doing grotesque cartwheels on the verge. We stop, as does the car behind us. 'Have you stopped for the bird?' Penny asks the woman, who is getting out of the driver's side of a bronze Skoda Fabia.

I recognize the type. If the car does not have an 'Atomic Power? No Thanks!' sticker it should. And so I rather like her. She introduces herself as Helen.

We walk down together to the crow, which is flapping, frantic. 'I love birds,' says Helen.

I wide-arm usher the crow into the hedge, so it cannot move, and then seize it. Clearly it has been hit by a car; the wing is dislocated, the eye bruised. It's young, a teenager. There's no point leaving it on the road, where it will be mangled more.

'Could take it to a vet,' suggests Helen. Then adds, laughing, 'And pay a fortune.'

'John's quite good at repairing animals,' ventures Penny.

And so we drive home with a juvenile carrion crow, held on my lap like a baby about to be presented to a vicar for christening.

To my relief the wing slips into place easily but the bruising is more severe than I guessed. The crow sags on one side, and is put to live in the protective custody of a dog crate, a hospital with bars. Feeding is tricky; the crow has to be held on its 'good' side, the jaw prised open, and food shoved in, six times a day, from a *carte* consisting of nuts, mealworm, seeds and blueberries. Especially blueberries. The edge of the beak has the paleness of youth. A bird's mouth inside is, I have forgotten, entirely odd: the tongue is hinged halfway along, with a hole in the base, for the trachea. And is surprisingly moist.

There is subtle beauty to the crow. While the principal colour is sheening black, there is a steely-blue lustre on the neck and back. I am amused that the crow's favourite food is the blueberry with which he tones so precisely.

After a week I can stop the force feeding, because he learns to open his mouth when tapped; his mouth agape is so large feeding is like chucking coins into a dustbin.

When I approach I call, 'Crow Crow.' I did consider a more in-tune, apposite 'Caw Caw' but when I say it, it sounds like something Sid James would say to Barbara Windsor. 'Corr. Corr.'

His greeting in response: a screech from *Jurassic Park*.

Four weeks later is Liberation Day. I put 'Crow Crow' (as he has become known) down on the lawn, and he manages a sort of glide a yard off the ground, like the Wright Brothers' plane at Kitty Hawk. So, it's a half-liberation. For a week he hangs around, flying higher and better.

The crows are few on our hill farm, so I take him down to Flinders, where crows abound. I spend a day with him, me doing office work in the back of the Land Rover (half of farming is form-filling), and then hiding food for him to find – nuts buried in the grass, blueberries in cracks of earth – as training in local scavenging. But best of all, he likes to sit on the wooden fence rail, waiting for me to bring him food. He still gets so excited he thrusts his head forward, and flaps his wings at my approach.

He watches other crows fly around.

That night, he stays at Flinders in his crate for safety, and next morning I let him out and feed him. He flies off happily.

For days Crow Crow, when he sees me, sails down and sits on the fence, and lets me stroke his chin while I feed him blueberries.

One day he disappears. I hope he has joined a murder – the collective noun for crows – and not become a meal for a predator. I never really find out, but either way it's a murder. One thing I do know: one of the greatest privileges in my life has been my bond with Crow Crow.

I would so like Crow Crow to come back to bite the hand that fed it, and eat the grain in the wheatfield. I think every crow I see is Crow Crow. Of course, in a sense every crow is.

The poets and pen-wielders have not given the crow much of a boost; on the contrary, they have execrated him:

My roost is the creaking gibbet's beam
Where the murderer's bones swing bleaching;
Where the clattering chain rings back again
To the night-wind's desolate screeching.

Eliza Cook, 'Song of the Carrion Crow', 1870

Lewis Carroll called the bird the 'monstrous crow, as black as a tar-barrel'.

Farmers have tended to give crows a blast of .12 bore shot, although as early as 1909 ornithologists Otto Herman and J. A. Owen in *Birds Useful and Harmful* (we had a copy at home when I was a child) argued:

The Carrion Crow follows the plough, and devours grubs and mice; it eats the insects in large quantities, and lies in wait, for the mice about their holes . . . The Carrion Crow steals and plunders the nests of the useful birds, spoils fruit and crops; but . . . these birds should not be too hastily destroyed, for they do

185

mischief only for a short time, while during the rest of the year they make war on the numerous pests, and are of great service to the husbandman.

Herman and Owen described the carrion crow as 'sly and cunning; courageous, but at the same time, cautious, and extraordinarily clever . . . Its sense of smell is very delicate; it scents carrion a mile away, under snow and earth.'

Certainly, Crow Crow could detect blueberries from a mile away.

14 JULY Flinders, drowning heat. The young starlings are standing panting in the grass margin, where the hares have eaten it down. Spiders are spinning webs between the ears of wheat, in expectation of the insects that come with fine weather.

In this breathless air, the birds have stopped singing; it's their summer recess, even for the robin, though the grasshoppers are in full compensatory chorus. In Aesop's fables a starving grasshopper begs food from an ant in winter. On being asked why he has not stored up food in the summer, the grasshopper explains: 'I had not leisure enough. I passed the days in singing.'

A common blue butterfly floats between the cornflowers, and there is sweet honeysuckle in the hedge.

Four of the red-legged partridge chicks have made it

to juvenile stage, and are already difficult to tell from their parents as the covey walks the side of the wheat, the cock bird having a real note of paternal pride in his stride.

Arcadia. For a moment. A woman in a black BMW 7 series then stops at the gate to ask how my 'rodents' are getting along. Rob, who is with me, says out of the corner of his mouth, 'BMW. Bling My Wheels.'

It takes a second to twig she means hares.

Hares, I explain, are not rodents but belong to the taxonomic order Lagomorpha, and differ from rodents in having four incisors in the upper jaw (not two, as in Rodentia), having a scrotum in front of the penis, and being wholly herbivorous.

I'm not sure which of us feels the more idiotic.

AN EVENING OF soothing calamine light. The wind wanders over the corn, which is full of the pollen and motes of summer.

A contemplation on hares, brought about by watching them at Flinders: putting hares back into the countryside is about more than 're-naturing'. Hares are one of the immemorial, defining English animals. When Vita Sackville-West wanted to invoke the unchanging pastoralism of England in *The Edwardians*, what natural items did she list? Well, 'the verdure of the trees, the hares and the deer'.

Have hares, have our national landscape.

*

LATE JULY Wheat does not go golden gradually, it does so in all of a rush. Two days ago I passed by Flinders and the wheat was still pale. Now the ears have trapped the colour of the blessed sun. Only when the wind ripples the crop can I see the green lingering down on the stalk. In a month the wheat will be harvestable. If I stand at the gate, Flinders looks as though someone has filled it with two foot of cream and then scattered confetti on top.

Breezes oscillate the standing wheat. Out of the depths a bumblebee climbs to the top of an ear, checks momentarily for direction, then sails away.

Bees have ever been part of the English scene. The bees flew when villa cornfields were scythed by Rome's woad-tattooed slaves. Bees hovered in the marsh flowers when Alfred's warriors formed the shield wall at Edington. The bees were there when Norman lords and ladies courted in the luxuriant flowery meads of castles, and, democratically, the bees were there in the back garden of a 1930s Durham miner tying up his runner beans.

More than mallets on croquet balls, the slap of oars on the river, or the chink of glasses of Pimm's, the murmur of bees is the sound of high summer in England. If one listens closely to the drone of bees one can hear something else: the hum of the Universe. Our historical closeness to bees is reflected in language. It was Chaucer

188

in 'The Squire's Tale' who coined the phrase 'as bizzy as a bee'. Then there is hive of industry, honeymoon, drone on, queen bee, bees' knees. British political philosophers have been tempted to look into apiarian society for Utopian analogies, as Bernard de Mandeville did in his *Fable of the Bees* of 1714. If de Mandeville saw proto-laissez-faire economics in the hive, others have glimpsed matriarchy, communism, although the surest quality of the bee is its industry. The bee is the Protestant work ethic on wings.

It is the unobtrusive things bees do for us that we are careless of. Who is the proclaimer of spring? Cuckoo? George Orwell, a keen naturalist, made an inspired case for *Bufo bufo* in 'Some Thoughts on the Common Toad', published in *Tribune* in 1946. But, no, it is the queen bumblebee emerging alone from her underground chamber in the hedge on that thankfully bright day back in February.

If the bumblebee entices the coming of spring, summer is truly the bee time. The level flight of bees from bloom to bloom sews the summer scene together. When a bee tumbles into a bloom the flower suddenly becomes vivid. This has happened now, with the bumblebee entering the flat purple eye of corncockle; I see that the petals are not uniform in colour but fade to light mauve in the centre. Corncockle is a handsome relative of the campions, pinks and ragged robin, and taller than the wheat it stands in couture contrast to.

And is there honey still for tea?

Not for much longer possibly.

The number of managed honey-bee colonies in the UK fell by 53 per cent between 1985 and 2005 and wild honey bees are thought to be nearly extinct throughout the British Isles. Reason No. 1? Agricultural pesticides. As the University of Reading reported in a major study on Britain's declining bees:

> Even when correctly applied, pesticides can have adverse impacts upon bees by reducing their breeding success and resistance to disease, and by reducing the availability of valuable forage plants. The report recommends that the government should commit to a targeted reduction in pesticide use by 2020. This should be accompanied by substantially improving pesticide labelling and accreditation regulations to mandate detailed assessments of the impacts upon all bees, not just honey bees.

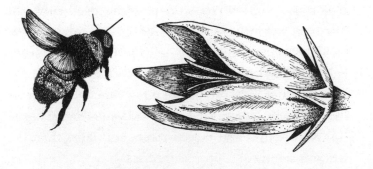

These are the same bees that agriculture relies on for pollination. The report also suggested that the decline in wildflowers was likely hastening the decline in bees, since bees and wildflowers go together . . . like bees and wildflowers.

Now for the madness of modern British agriculture, its bonfire of the sanities. As research published in *Ecology and Evolution* and the *Journal of Applied Ecology* has proved, soft fruit production, yield and value can be increased by 54 per cent just by having an adjacent wild-flower strip to encourage and feed bees.

British bees are adapted for British native flowers. They really are. All over the wheatfield there are elegant eruptions of red, blue, purple and yellow amid the green wheat. The borders of the field are a mass of poppies, corn chamomile, corn marigold and cornflowers, and self-seeded indigo wild pansies and scarlet pimpernel. In the drowning July heat I can see red carder bees, honey bees, red-tailed bees. There's a bee I can never identify, so I've brought my DK insect guide and a jam jar. I trap the bee under the jam jar; it is a buzzing torture for it to be imprisoned, and a buzzing headache for me to identify it. Finally: it is a field cuckoo bumblebee. When I release it, the bee attacks me, which is fair.

In memoriam: twenty-three species of bees and flower-visiting wasps have become extinct in the last 160 years.

TIME DOES NOT erode; it builds sedimental layers of particulate memory. And is it really true that familiarity breeds contempt? Doesn't it just engender a deeper love? We are in a running night tide of cars, the town names of mid-Wales (Penybont, Rhayader, Capel Bangor) flashing up like navigational buoys in the headlights – though I hardly need guidance, because I have done this journey as boy and man so many times I have lost count. There's the place my father's custard-yellow Rover 2000 broke down, and we had to get stream-water for the radiator; there's the bend where I saw a pine marten . . . We are driving to Borth on Cardigan Bay, because that's where Herefordians go to the seaside. Always. We land at Borth's top end, slide open the windows so that the dawn-creech of the herring gulls and the ionizing air of the Irish Sea banishes the car-fug; then we drift along the town's improbably long one-horse street, playing the mirthful game: what's changed in Borth since last time? Nothing! It's the town preserved in brine. Out beyond the lazy, can-hardly-be-bothered breakers, gannets plunge into water the tone of mercury. On past the fishermen's cottages, past the mirey mysterious Cors Fochno marsh to Ynyslas dunes, where we park, herringbone style, on the hard sand of the estuary, the Dyfi river a faraway trickle across the expanse of sand. The Land Rover next to ours has a windscreen sticker for Evans Farm Supplies on Homer Road; the same merchants we use. On the other side is the plumber from Bacton. Half of Herefordshire might be here, but the towering dunes, which bristle with marram grass, are so vast that we escape through the valleys into a private world, cut off from work cares by the sea haze which means anything beyond 20 metres is out of focus . . .

I see myself age nine (wearing a rollneck jumper; it was the pre-fleece 1970s) gathering smoothed driftwood on the straight mile of shoreline, then spotting a coloured art-deco brooch. But it wasn't a treasure made by man; scratching away the storm-dumped twigs and grit, I uncovered the beak and then the body of a puffin. There was not a mark on its innocent white chest. I had finally laid eyes on a puffin – and it was dead. (Yes, thinking back, that was the moment I became an environmentalist.) The memory sticks on the retina. Then again, you need teary salt-tang to better appreciate the sweetness of life . . .

Now I am thirty-nine, and we are jumping off the top of the dunes into sandpits spooned out by the wind. Tris and I are Action-Man flying, Freda fairy-gliding. Our sole companions are the oystercatchers piping on the liminal shingle in front of us, in the exact place where I found the corpse of the clown-bird. Freda has fits of giggles because her miniature Jack Russell keeps pooing, and she has to dispose of the evidence with her gift-shop spade, which really isn't quite long enough. A flicker of a squall only creates grey bullet holes in the golden sand for us to dodge. Penny, wearing a retro rollneck sweater, is down in the dune's base and content because she has found a wind-free space for a Primus to blue-burn to make a cup of tea. Ah, but time is running out; we have to catch the tide home. There is one more ritual to observe, however, when we get back in the Land Rover.

Years ago: I sat excited on my father's lap to drive on the compact, wet sand of the estuary. Two years ago, Tris sat on my lap and drove. Last year, Freda sat on my lap and drove. This year, Freda sits on Tris who sits on me. Layer upon layer.

There is a familiar level of stink in the car too, above the smell of wet dog. Loaded in the back are six sacks of seaweed we have collected. It's mostly drift bladder-wrack and kelp, collected from mid-shore – there is no point collecting wet seaweed, with its excess weight and salt. There is a long history, back to Roman times, of British coastal people using seaweeds, especially the large brown seaweeds, to fertilize land. If you wander the coast of Wales you'll sometimes come across an abandoned, scrub-covered access ramp at the back of the beach from when the seaweed used to be carted off by horse. Often seaweeds and kelps contain high amounts of potash. I don't know when we started using seaweed as a manure, I just know that we always have. I suspect we got the idea from one of the agricultural salesmen who used to come to the farms and hawk a commercial seaweed extract called Maxicrop.

In Herefordshire that would be a lost cause. Why pay good money for something you could collect at Borth after you had built the stone windbreak?

So, we always came home with bags of seaweed, a little of which, dug into the manure heaps from the stables, the cow sheds, the chicken sheds, went a long way.

Poppop and Grandma also strung up lengths of black bladderwrack next to the barometer in the hall in an attempt to divine the weather. Sometimes in checking the weather they fell into outright superstition and looked in the newspaper for the Met Office's prognosis.

Also, the trip to the seaside was – or indeed is – incomplete without gathering a bucket of oyster shells, to be crushed at home and given to the chickens. Nothing makes better eggshells than seashells.

The children are ten years older, teenagers, and harder to round up, but we still make it to Borth. And we come back every time with the seaweed and the oyster shells.

THE WHEAT, ORIENT and golden. A slight, south wind.

As I walk the edge of the golden sea, the crop – almost ripe now – is eurythmic. Then comes an explosion of grey shot, as six partridge rocket out of the wheat, flying wildly away, cackling as they go. Another shoots up and after them on whirring wings.

Appropriately enough, 'partridge' is derived from the Greek *perdesthai*, which means 'to make explosive noises'.

From behind, at speed, it can be hard to tell

red-legged partridges (*Alectoris rufa rufa*) from grey partridges (*Perdix perdix*), but on this sun-happy, clear-skied morning I can plainly see the white throat, black necklace and ruby beak of the red. The red-legs are homely creatures and, despite being put up (accidentally) by Edith, if food is abundant they will not wander far.

That night I go to the field, which is restless under the moonlight. At the edge of the wheat the red-legs are hunkered down. Like pheasants, they prefer to become sculpture-immaculate still when danger is about; it's motion which most often gives the gamebird away to the predator. My torch has no difficulty spotting them.

The red-legs are in a circle, their heads pointing out, so enemies can be easily spotted. This habit of roosting is known as 'jugging', although if the wind is up all the covey will face into it.

But I push my luck too far, go too close, and they melt away into the wheat.

The next morning there is a perfect ring of black droppings where the 'jug' had been.

SOMEWHERE UP THE M6 on a Saturday at 4.59pm, round about Birmingham. I switch over the car radio from R3 to R4 to catch *PM*. It must be a slow news day because headline number three is 'Nature writer attacks BBC *Springwatch*: Chris Packham hits back'.

They can only mean me. I switch off the radio and, after stifling a cardiac arrest, exit at the next services and phone Penny.

She agrees to listen. I can't face it.

Two hours towards Scotland later, I phone Penny on my mobile. 'It was fine,' she says. 'Really he was left agreeing with you.'

I had guessed what it was about. In a book called *Meadowland* I had wondered idly whether repeated images of an animal on a screen devalues the experience of encountering said animal in the wild. Hardly original, I know; I borrowed the essential idea from the Marxist cultural critic Walter Benjamin and his *The Work of Art in the Age of Mechanical Reproduction*. I'd mused equally idly on the same notion when talking at the Hay-on-Wye literature festival on what was a slow news day for journalist Hannah Furness, who put the iconoclastic claim on the pages of the *Daily Telegraph*.

Do I not have a point though? If, as all allow, watching violent video games coarsens sensitivity, does not watching endless nature programmes coarsen wonder? Reduce nature to spectacle, to entertainment?

Obviously, if one wanted to flag up *Springwatch*'s irredeemable flaw it would be this: the programme format requires a constant stream of images of charismatic animals. No viewer wants to sit watching a screen where nothing happens. But in truth, if all the *Springwatch/ Autumnwatch* cameras were pointed at the average

197

agricultural field in Britain the screen would be an unchanging image for two hours. There is no wildlife to goo over. Nothing stirs, nothing moves. A still photograph held in front of the camera lens would serve.

Dangerously, *Springwatch* suggests nature in Britain is busy and bountiful. It is not.

Fortunately, it seems that today's *PM* is somewhat low in listeners, and I do not get anything like the Twitter trolling I expected for criticizing a national institution.

LATE EVENING AT Flinders: under a strange sky made from a witch's scarlet cauldron-flames, the partridge come express-sprinting down the paddock, so smooth in their movement they seem to be on rollers. One makes a rally-ing call as it slips through the wooden field gate into Flinders, and leads the covey for the nearest row of wheat. The partridges' *go-chuck* calls are becoming the sound of Flinders' evenings.

The edge of darkness now: the partridge hunt around the wheat; then a sparrowhawk veers in to hunt them. Two carrion crows are ripping the top off a stalk, shredding it, plucking it. I am glad of their vandalism this evening, since on seeing the sparrowhawk they turn vigilante and lurch up to drive it away.

The red-legs live to peck and to pleasure another day.

WE START OFF under a red morning sky, so there is no lie there. By Leicestershire it is pouring with rain and the land getting flatter, and by Cambridgeshire it is flat. By Norfolk a sausage would not roll.

It is decades since I have been to East Anglia, but there is a strong incentive. We need a new Red Poll bull, and a likely contender is advertised on a livestock website. I'm mesmerized by his stats. He is a begetter of descendants to rival an Old Testament patriarch. Also, Red Polls are a Norfolk/Suffolk breed so there is appositeness in purchasing one from its place of origin. Coal from Newcastle; sand from the Sahara; Red Polls from near Swaffham. It turns out, however, that when talking to the seller on the phone I should have been more inquisitorial over the phrases 'used to being handled' and 'been on the halter'.

Somewhere near Whittlesey, after three hours' journey, Penny and I swap the driving, which is a good idea because I'm driftingly distracted by the size of the fields as we go further and further into the realm of the grain barons.

We are passing wheatfields of a hundred acres or more, greater than all our land and fields put together. Since the cutting season is earlier in the dry east most fields have been harvested, and there are yellow ziggurats of square bales to the horizon. One field is still being harvested, with three combines side by side.

Eventually Mrs Sat Nav tells us, 'You have reached your destination on the right': a picturesque red-brick farmhouse, red-brick barns, all perfectly colour coordinated with Red Poll cattle. The seller and his teenage son come out of the Georgian door, which is heavy with brassware, on to the gravel. There is the shake of hands, the usual small talk.

'Did it take you long from Hertfordshire?'

'Herefordshire.'

We walk around to the square yard behind the house. A flock of paper-white seagulls circles overhead. The rain has stopped, though the afternoon sky has taken on the tone of dead flesh. The bull is in a corner pen with concrete walls topped by horizontal steel rails, hand-thick. There is a feeding and watering hatch. Also a yoke, so the bull can be restrained by the handler outside the pen.

The bull is magnificent, his glossy conker coat a halo around him. The trouble is his temperament. The moment he sees us, he lowers his head and 'makes eyes'. He bulges them out so the whites are a visible ring. This is the prelude to action.

The bull does not know me, so a certain wariness is to be expected. But then I suggest to the seller that we all withdraw and he approaches the bull while I watch from a distance. The air becomes brittle with embarrassment.

In the yard entrance, I stand with the son. Out of the corner of my eye I see he is looking down at the ground. The seller does not get to within twenty feet before the bull drops his head. This time, he snorts.

I give the bull one more chance. The seller and his son wait by the yard entrance and I walk up to the pen slowly, smiling, calling, cooing. He allows me within ten feet before he suddenly shifts through the threat levels to critical: attack imminent. He starts pawing the floor with his right front hoof. I can see into the cowshed from here: there is an elaborate system of races (corridors of steel bars) and gates from the outside pen into the shed and yard behind. The bull, I suspect, is kept separate from the herd except when procreation is required.

I doubt if he is handled at all, except remotely with yokes and crushes. Or been on the halter since he was a calf. He is caged, like King Kong was caged. It is not unusual down on the modern farm, this remote-control of livestock, where food and bedding is delivered by a blower on the back of a tractor. All is fine and dandy, except when you have to get in with animals who are unused to handling, to the close proximity of humans. And sometimes you do have to get in with the stock.

No bull is risk-free. Being with a bull on a farm is exactly like being a runner at Pamplona. You should always have an eye on a safe exit. Nevertheless, there are degrees of danger, and this bull is over my personal line. The old bull who sires our calves you can lead by the nose. Really. You put your fingers through his punky copper nose-ring. If he gets naughty, the ring gets a tiny twist: he sighs, and behaves himself, and walks like an overgrown dog.

I am somewhat annoyed by this bull-buying failure, by my credulity over the advert, by the four-and-a-half-hour drive, by their lack of honesty.

'He's not for me, I'm afraid,' I say.

'We've wasted our time,' they say, assuming the poise of injured party.

BUT OUR DAY IS not entirely wasted. An hour later, Penny and I are sitting in a remote car park at the end of a lane in Coates. There are a couple of pick-ups, a couple of saloon cars. Two guys in green macs are taking photographs with cameras with big lenses.

The RSPB's reserve at Nene Washes is the site of a corncrake-reintroduction scheme, begun in 2003. Once, the corncrake was a widespread bird of hay meadows and arable fields, but it could not cope with the mechanization of mowing. The corncrake had the suicidal impulse of many camouflaged ground-nesters to sit tight, even as the cutters made their inexorable approach. By the Edwardian era, the bird had already become very scarce in southern England and the Midlands. Joseph Whitaker, the Nottinghamshire bird-man, wrote in *Jottings of a Naturalist*, 1912:

Twenty years ago there were Corncrakes all over the parish, in fact it was the exception not to hear them in every mowing field, but I know that there has been none for the last ten years, not a single bird heard, and the parish is six thousand acres; and it is not only so in these parts, it is the same everywhere.

The decline of the corncrake, the 'landrail', the 'darker-hen', continued through the twentieth century until by the 1990s corncrakes were restricted almost entirely to the islands on the north and west coasts of Scotland, where a less intensive form of agriculture – crofting – allowed the bird to retain a clawhold in Britain.

The reintroduction project at the Nene Washes involves releasing hand-reared corncrakes, bred at Whipsnade Zoo. Corncrakes are only summer visitors to the UK; in autumn the birds migrate to central Africa to spend the winter. Corncrakes fly slowly, with their legs dangling, every beat of their wings effortful. It is a miracle that any corncrake can fly to and from Africa, particularly birds raised in zoological gardens. But they do.

There are now about twenty male corncrakes on the Nene Washes, some of them born wild, with established territories.

The Nene Washes are a six-mile-long oblong of horizontal, raggedy, marshy, table-flat grassland that store floodwater from the River Nene. The elements make for the best conservators: the grassland has never been improved because it is inundated in winter.

We walk along the path by the dyke, properly called Morton's Leam after the bishop who arranged its excavation in 1478. At the end of the long, exhilaratingly flat view, as far as the eye can see, the point at which the grass line disappears into the sky line, Hercules aircraft pass. This is still, among all its uses, RAF bomber country.

Reeds and rushes wave. Across the ink line of the leam, Limousin cattle graze. A marsh harrier hovers, and shoveler ducks rise, panicking, in spumes of spray. I see kestrels, tree sparrows, snipe. A minute scarlet javelin on top of a fence post catches my eye; a red damselfly.

We walk for a mile or more. No corncrake calls. There is only the lackadaisical wind, the endless mauve sky, transparent disappointment.

I so, so badly wanted to hear a corncrake again.

We are actually back at the car park when a man, all camouflage kit and ginger beard, hisses at two women beside a Ford Mondeo, 'Listen!'

I go through the various stages of disbelief. Mishearing. The passing thought that I am dreaming. That I am inside a Wordsworthian relapse caused by desire:

> There are in our existence spots of time,
> That with distinct pre-eminence retain . . .

But there it is, real, audible, with four other human witnesses. *Crex, crex*. A sound like no other. It's *something* like a fingernail on a comb; it is a *bit* similar to a washboard in a skiffle band. A corncrake is calling to me.

The years flash back, through my childhood, through Victorian industry to Georgian times, when corncrakes were the sound of summer in England. I hear the wonderful, bird-bountiful past.

The bird crakes again, over and over, from within the grass vastness beyond the leam.

In this moment on the Nene Washes I understand that time is not linear. In the calling of the corncrake the loops of past, present and future are pinched together between finger and thumb.

It is getting on for night when we leave. The corncrake is still calling.

WE SPEND THE NIGHT at the William Cecil Hotel in Stamford, which takes dogs if they are 'reasonably well behaved'. Thus Edith the Labrador qualifies. It is the sort of hotel where they still have keys, rather than plastic cards in slots.

There's also a certain amount of historical nosiness in staying at this hotel. William Cecil's family came from my parish, and he ended up being ennobled by Queen Elizabeth as Lord Burghley. The hotel is part of the Burghley House estate. William Cecil is the original local lad who did well.

My nosiness extends to eavesdropping on the table next to us at dinner, which consists of five businessmen, all of them in black shirts, most of them with gelled hair. Three are English, two Dutch. They are in the flower business. Apparently the market for bulbs is dropping because gardens are getting smaller.

The figures for this end of agriculture are stupendous. 'Billion-Euro turnover . . .' It's all pallets, refrigeration, Tesco, aggressive marketing ('He called to say, "Don't fuck with me"'). The fate of Manchester United is a conversational emollient. What to do about the dominance of the Dutch Flower Group is the main of the conversation.

*

How sweet and pleasant grows the way
Through summer time again
While Landrails call from day to day
Amid the grass and grain

We hear it in the weeding time
When knee deep waves the corn
We hear it in the summers prime
Through meadows night and morn

And now I hear it in the grass
That grows as sweet again
And let a minutes notice pass
And now tis in the grain

Tis like a fancy everywhere
A sort of living doubt
We know tis something but it neer
Will blab the secret out

If heard in close or meadow plots
It flies if we pursue
But follows if we notice not
The close and meadow through

Boys know the note of many a bird
In their birdnesting bounds
But when the landrails noise is heard
They wonder at the sounds

They look in every tuft of grass
Thats in their rambles met
They peep in every bush they pass
And none the wiser get

And still they hear the craiking sound
And still they wonder why
It surely cant be under ground
Nor is it in the sky

And yet tis heard in every vale
An undiscovered song
And makes a pleasant wonder tale
For all the summer long

The shepherd whistles through his hands
And starts with many a whoop
His busy dog across the lands
In hopes to fright it up

Tis still a minutes length or more
Till dogs are off and gone
Then sings and louder than before
But keeps the secret on

Yet accident will often meet
The nest within its way
And weeders when they weed the wheat
Discover where they lay

And mowers on the meadow lea
Chance on their noisy guest
And wonder what the bird can be
That lays without a nest

In simple holes that birds will rake
When dusting on the ground
They drop their eggs of curious make
Deep blotched and nearly round

A mystery still to men and boys
Who know not where they lay
And guess it but a summer noise
Among the meadow hay.

John Clare, 'The Landrail'

*

THE VILLAGE OF Helpston is only seven miles from Stamford. So we go next morning, on a sort of pilgrimage to John Clare.

Helpston is a vision of a village, with its honey-stone vicarage. It is quiet, of course it is, because its inhabitants are away working, the drives empty of cars.

There's a brown heritage sign for 'John Clare's Cottage' on Woodgate Lane. It is where he was born in 1793. Whitewashed, with a thatched roof, it is now a museum. Standing in front, one thinks, 'Quite big for a farm labourer's house,' before noticing the little bricked-up doorway. Originally it was sub-divided.

There is little in Clare's background or early childhood to explain how he became the greatest poet of the British countryside. He himself said he 'found' his poems in the fields. He may well have been telling the truth. If ever nature possessed a human in order to have a voice, then Clare was the chosen one.

Clare's cottage does nothing for me. Why would it? Clare was the outside outsider. I go for a walk past the farm buildings opposite, where both he and his father worked, following the public footpath. There are signs on the black corrugated sheds warning of private property and the consequences of trespass.

Beyond the sheds the fields start. Then it hits. No child could now run, as Clare did, into 'the corn to get the red and blue flowers for cockades to play at soldiers', because there are no arable flowers in the fields, which are maddeningly oversized, dreary and dead.

Pylons march across the land. No birds sing, let alone landrails.

FLINDERS: THE SOUND of the wind in the wheat is the sound of falling sand; the swaying of the wheat gives body to the invisible wind. I'm splitting grains with my thumbnail to determine ripeness and readiness (if it's 'milky' in texture it's not ripe), when a nightingale starts singing in Three Acre Wood. I'm lured into the dark trees by the siren songster, like a naïve child in a German fairy tale. Deeper and deeper into the wood I go, but the nightingale is always a little ahead, hidden by leaves to preserve its modesty/mystique; the leaves are its harem screen. Or perhaps the bird is ashamed of its drabness; the nightingale is a plain Jane/John. At one point I stand under an ash and the bird is singing above me, the scales of its song tumbling on to my upturned face.

The wheat is weeks off readiness.

WE DRIVE FURTHER than other people on holiday.

At the tolls on the autostrada north of Milan, the guy in the stainless-steel kiosk (the Italians can style-up everything, even toll booths) hands down the change to Penny and says, 'First English car.'

'Today?'

He ponders for a second, and says, 'This year.'

He may not be exaggerating. We last saw a GB plate back in Belgium, and children in overtaking cars point at us excitedly. You don't need to be able to lip-read Italian; their faces say it plain enough: 'Look! English people!'

Sergio, as the Agip pump attendant at Bergamo services introduces himself, can't believe his luck, for we are an opportunity for him to practise his English.

'Welcome in Italy! Or is it "Welcome to Italy"?'

'To,' I inform him.

'You from Liverpool? The Beatles! London? Freddie Mercury!' (This one takes me a minute; it comes off Sergio's tongue as 'Freedie Muuur-coo-ree'.)

I shake my head. Sergio knows only one other place in Britain: Oxford. We settle on close to Oxford because, after all, what is sixty miles between new-found friends.

I like driving abroad. I like seeing the countryside change, I like seeing how the other 90 per cent farm, and the roads are fast and relatively traffic-free.

Autostrada A4 Torino–Trieste is an exception. Two of the three lanes are solid with lorries. We don't get much further along before the traffic locks, and we go down to crawling pace.

The Saab's computer reads 42 degrees outside temperature. Beyond the windows the scene is undulating, belly-dancing in the heat. Beside the autostrada there are factories, car showrooms, industrial parks. Occasionally some farmland. We creep past a low, white two-storey farmhouse, level with a wheatfield. The heat and the dust of midday; factory-scale noise of cicadas in trees. (Adult male cicadas possess two ribbed membranes called tymbals, one on each side of their first abdominal segment. By contracting the tymbal muscle, the cicada buckles the membrane inward, producing a loud click. As the membrane snaps back, it clicks again. The two tymbals click alternately. Air sacs in the hollow abdominal cavity amplify the clicking sounds. The vibration travels through the body to tympani which amplify the sound further. A single male cicada can make a noise over a hundred decibels. Just so you know.) Through the haze, behind the wire-netting fence, I see a blue Ford tractor coming towards me; the tractor has a shiny new rear-mounted attachment which is cutting wheat and binding it into sheaves – modern tech doing the old ways. It is my road-to-Damascus moment. I want to get out for a closer look, but the traffic suddenly frees, and we're off. I have to wait until Treporti to Google ('modern tractor mounted reaper binder') this wondrous contraption. It is an Alvan Blanch TH1400 tractor-mounted harvester, and it turns out to be made in England, in Wiltshire.

I need one, either bought, begged or borrowed. My intention had been to buy a vintage reaper-binder (you can still pick them up at farm sales), which can only be described as a spindly steamer paddle attached to a child's go-cart. They are notoriously temperamental, and require the patience of St Monica.

I do mean I need an Alvan Blanch TH1400. I am not getting my wants and needs confused.

31 JULY Flinders. How swiftly summer comes. And goes. Over the hill, one of the farms is already cutting wheat. Although it seems only yesterday, it is really more than four months since standing in the field at night I watched the lark ascending.

CHAPTER V

Of Men and Harvest Mice

Anon the fields are wearing clear
And glad sounds hum in labour's ear
When childern halo 'here they come'
And run to meet the harvest home
Stuck thick with boughs and thronged with boys
Who mingle loud a merry noise
Glad that the harvest's end is nigh
And weary labour nearly bye
Where when they meet the stack thronged yard
Cross bunns or pence their shouts reward.

John Clare, *The Shepherd's Calendar*

I PHONE THE NICE sales rep at Alvan Blanch, who gives me the price of their reaper-binder wonder gadget: £7,642 including 5 per cent single purchase discount. Joy unconfined. Although, by the measure of agricultural expenditure, this is a very reasonable price, I cannot justify £7,642 on something that may only be used once.

Eventually, I find someone I can rent one from by the week. This is David Baldwin, who farms outside Chard in Somerset. Only problem is, I have to collect it.

I love our Land Rover Defender, which is as much fun as you are likely to have with four wheels. But the noise and vibrations kill.

One example will suffice: we usually have a sensible northern European estate car for family journeys, but once said northern European car inconveniently broke down just before my nephew's wedding, requiring us to drive a hundred miles to Basingstoke in the Land Rover, three of us in the cab, four-year-old daughter's car seat tied (and by tied, I mean with nylon blue ropes) to some hurriedly inserted bolts in the 'utility space' normally occupied by sheep/fodder/kit. Just to add to the tension, while I had been cleaning my teeth before set-off I had looked out of the bathroom window – to see our herd of Gloucester Old Spot pigs perkily trotting along the yard on a day out.

After swearing more copiously than Hugh Grant in the opening scene of *Four Weddings and a Funeral* as he tries to get to the church on time (twelve F***s, by my count), I changed back into the ubiquitous boiler suit, and eventually enticed the pigs into their rightful paddock with a year's worth of please, please, please sow nuts.

Since we were two hours late in starting off, I had to drive the Land Rover Defender outside its envelope, which is 60mph. At Leigh Delamere services we stopped, and as we were clambering out, a white-haired man, crooked with age, helped by his son, tottered up to us.

I recognized him instantly. He was my dead grandfather. You can tell old farmers; they are thin men made from steel hawsers.

'She was burning clean,' he said to me, pointing at the Land Rover. 'We've been following her for a while.'

We shook hands, chatted, he asked, 'Far to go?'

'About seventy miles.'

He winked because he knew: 'Luck with that.'

We got to the country-house wedding venue, parked up at 1pm, just in time for the service. And that is all I can remember till about seven in the evening and seeing my tiny daughter dancing in a blue velvet dress. The juddering from the engine, the drenching struggle to hold the Land Rover on the road, had left me a physical, traumatized wreck. People talk about ringing in the ears: I had ringing in my entire body.

I try to bury this memory on 10 August when I set off for Chard with a trailer on the back of the Land Rover: I reach Chard, load the Alvan reaper-binder, drive home. Everything else about the day is erased due to amnesia caused by mechanical shaking.

Since the Ferguson is a bit light and underpowered for the Alvan reaper-binder I hitch the cutter to the back of the International and drive over to Flinders. When I get there it is thundering and wind is whipping the wheat. Young swifts, fresh to flying, hawk for flies over crashing golden waves.

I AM GULLIVER in Lilliput, bending down to part the stalks (now sun-polished and ramrod straight). There is squeaking in the wheat; not unpleasant; rather more euphoniously soprano. I keep parting the bambooey stems, down to the earth.

Two field mice, impenetrably black-eyed; they look at me, dismiss me. Return to their own time.

You think of mice, and you think chill, needle-claw scuttling. These two mice are plump and hot and brown-silky. They do 'tag' around and up, down the stalks; then they do hide-and-seek. An animal behaviourist would doubtless reduce the motions of these mice to either kid-learning or adult pre-fucking.

I'll tell you what it looks like. It looks like fun.

The thick heat of the wheatfield at soil level is intoxicating. It smells of a newborn baby's head, or a loaf fresh from the oven.

I'M AT FLINDERS every day now, checking the grain, rolling an ear between my palms, then splitting it between thumb and forefinger.

By 20 August we are ready to harvest. The grain splits hard, like a nut. (You can use an electronic moisture meter; grain needs to have a moisture content below 16 per cent, otherwise it will rot.) Nature only gives a couple of days between the crop that is ideally ripe, and one which is deteriorating. Or, worst of all, sprouting in the ear, which is what did for Henchard, corn merchant and Mayor of Casterbridge.

I'll be cutting the wheat tomorrow, and am suddenly struck by a terrible nostalgia. This golden sea with its daubs and borders of flowers will be gone. For ever.

I lie down in the wheat, so I might know the world in which the hares and the red-legs live.

My God: the terror and the beauty.

The robin in the hedge begins a wistful song. Robins are singing again because they have finished moulting and are taking up their winter lands. They are unusual birds because both male and female sing and hold territories.

The docks in the field margin are rusting; chains of bryony, Tic-Tac orange and lime green, are tying up the hedges, there are insistent wasps on the flowers of the ivy, and daddy-long-legs are puppets on invisible strings. The evening swirl of swifts has become thick and black, because the young birds have joined in.

Summer is past its peak.

IF I HAD A BRAND-NEW combine harvester I'd give you the key. Because I'd rather not have it myself. Aside from wishing, due to ecological desires, to cut my wheat into sheaths rather than bales, there is a personal reason why I avoid combines.

I think it was my sixth birthday, and my father gave me a toy farmyard he had made. The idea was brilliantly simple: a two-foot square of thin plywood, covered in wood glue, sprinkled with sawdust, then covered with green paint. A woman's round handbag mirror, stuck in a corner, served as the farm pond. Fences were batons with pegs, around which my father had woven twigs in imitation of hazel hurdles.

I was always mad about farming, but it is only now, looking back four decades, that I realize just how consciously I reproduced on that board the world around me. There were the plastic Red Poll cattle, facsimiles of the herd along from the village school, the brilliant white David Brown Selectamatic tractor which my grandfather drove, the flock of white geese (miniatures of ours), the sheepdog (ditto) and the Dinky version of a Massey 575 combine harvester. Uncle Ron had a pillar-box-red Massey 575.

I didn't play with the Dinky combine very much, an occasional backwards-forwards swiping, while everything else was tended with the care of a particularly saintly nurse. I hated Uncle Ron's combine. For eleven months of the year it was imprisoned in the barn; the stone walls and heavy barn door barely restrained its malevolence.

Come July, the Massey was let out for its annual mayhem. True, Uncle Ron's Massey never actually killed anyone, but in this it was almost unique. For years as a child I thought the phrase 'tragic accident' was a synonym for death by combine harvester. Our neighbour at Withington, Bill Williamson, was eaten alive by his Massey; a boy at Bromyard had his arm torn off; a girl at Stoke Edith 'lost' her hand.

Skip forward: I'm twelve, and it's midday and high

summer, although one cannot tell this, because the dust from the combine's cutter sends up a grey cloud; we are in an eclipse, in constant twilight.

We're cutting wheat on the top field of Uncle Ron's farm at Upper Sapey. Everyone over twelve (me now!) is doing something, because tractors with trailers have to be constantly pulled up alongside the combine to take away the grain from the unloader pipe.

Well, everyone except Benji, who is ten; he's got his Webley .177 air rifle to ping rats as they flee the cutter.

My job is to stand on the small open balcony of the Massey, next to Olly, and signal to the tractor driver alongside when his trailer is full.

It's terrifying being on the balcony; it's like being on the tiny bridge of a battleship as it pounds through waves. The balcony, with its two low, pathetic rails, is perched above the cutters, so one always feels as if one is falling on to them, particularly when the combine hits a ridge and pitches forward.

The pandemonium noise, the perpetual whirlwind of straw fragments and earth. Above the cacophony, Olly shouts, 'Do you want a go?' I can't really hear the words, only follow the outline of his lips, now he has taken the red Wild West handkerchief away from his mouth. He's wearing Second World War fighter-pilot goggles, found in a second-hand shop in Tenbury Wells.

I slip into the seat; I've got sunglasses on as protection. Fighter-pilot goggles or Polaroids – such is 'safety equipment' in the 1970s. I can barely hold the 575 on the line, but peer pressure is the devil's whip, and I have to keep going to the end of the field at least. The field is long, longer than any prairie. Olly is swigging bright-red pop from a bottle. Coca-Cola, Vimto, Cherryade – it's all pop to us.

My leg trembles with the pain of keeping the accelerator down. And why don't animals move out of the way? A rabbit is flung up and back down into the blades, throwing a bouquet of scarlet against the artificial thundercloud.

After one run down the Canadian vastness, Olly takes over again, so at least I am not driving when it happens.

The terriers have come up with us, because there is nothing they like better than to chase the rats bolting out of the wheat. One grey rat decides to do a scurrying U-turn, back towards the blades. And Bella the tan-and-white terrier is right on its scaly tail.

There's a frantic shout, from me, from Olly: 'Bella!'

But Bella never hears, not above that industrial din. Her head is cut off at the neck, and flies up towards us, her eyes wide in surprise. I remember being sick. I remember Olly banging the rail in despair. Bella is my cousin Alys's dog; someone will have to tell her when we return to the farm. We do not stop, we do not turn the Massey off, because it will probably not start again. We are pretty much baling as we go, and it is on the next-but-one run along the field that I see the straw bale with bits of brown and white fur, and the butcher's ochre of dead flesh.

On that yellow-remembered hill there was trouble

still to come. I was fourteen, and standing on the Massey's balcony about to drive when thick, nidorous smoke slid out from the engine, on a register above the usual throat-catching stink of hot oil.

Orange flames blossomed at my feet.

We jumped for it. For an hour the gigantic Massey burned, as though it was the effigy in some pagan ritual, or the denouement of a science-fiction movie of its own making.

IT IS A WEEK AFTER the Massey combine incinerated. Late morning. We've not had elevenses, I remember that. The baler's broken down, and everyone is on the yard. Jack R, the mechanic who works for my father, is on his back underneath it, blue metal toolbox with its concertina trays open. Slicked-down wavy grey hair, pugilist's face, Jack could do a sideline as a character actor (speciality: nose-tapping villains, who say 'Know what I mean, son?').

I like Jack, and not only because he has bought me a present at Christmas for as long as I can remember. He's called Jack R, because everyone middle-aged in the 1970s was called Jack. So he has to be Jack R(eynolds) to distinguish him from Jack Williams, Little Jack, Jack Cooper, Jack-Down-the-Hill, and Leysters Jack.

Jack R, in his habitual blue nylon overall, is muttering. Usually, watching oily-handed Jack is interesting, but I can't see what he is doing. It is at this point that I have my Big Idea.

Almost all kids in the country in the 1970s could ride a horse to some degree, although of my friends and family I only remember my cousin Alys and I having lessons. At my riding school at New Court on the Lugg Meadows I've been having polo lessons, although this is a rather grand term for what largely consisted of sitting, legs astride, on a raised wooden tub and swinging a polo stick. Nonetheless I have now played in two polo matches and have a polo shirt and wrist sweatbands, so am expert and Know Everything.

To assembled cousins on the yard I say, 'Why don't we have a game of polo?' They all look appropriately nonplussed, but I press ahead.

At some level, I must know this is a poor notion because I say it sotto voce, beneath the level of adult hearing.

Since we are at the Upper Sapey farm we have horses. Lots of horses, because Uncle Ron has hunters and racehorses, and my cousins have ponies. We dismiss the racehorses because sitting on a racehorse is akin to getting in an F1 car with a nitrous-oxide booster, but faster by a factor of three. (You haven't experienced fear until you've ridden a racehorse; I'd just 'ridden out' on one.) This leaves three hunters and three ponies. Saddled up, with hockey sticks and upended walking sticks in lieu of polo mallets, we ride to the paddock behind the new steel-frame cowshed.

The goalposts, of course, are jumpers. Since the walking/hockey sticks are far too short to lean over with to hit the ball – even a football – two younger

223

cousins fall off instantly, cry, and don't want to play any more. With me on Foxley, my uncle's 16.3 Irish hunter, and a just-remembered shepherd's crook as a mallet we have some fun for ten minutes.

We are playing the second chukka when I wang the crook wildly, and hit Foxley a terrible blow to his front right flexor tendon. The horse shies, lands, stumbles.

There is O-mouthed silence. I'm out of the saddle and running my hand over the tendon, and can feel it actually swelling under my hand.

Cousins are standing behind me; someone says, half sympathetic, half gleeful, 'He's going to kill you.'

Uncle Ron has just the fiery temper his red-raw face suggests, so execution is not exactly off the tariff of punishment. Foxley is his pride and joy and, since money is everything, the horse's financial worth must be stated. Foxley is worth a thousand guineas. I know this because everyone says rhetorically, with a rising high pitch, 'Do you know how much 'e paid for that chestnut gelding? A thousand guineas!'

Dilemma. Do I confess? The damage to Foxley could be permanent. It could be a chipped bone, a severed tendon . . . At this point, Alys, just nine but with the wisdom of an ancient, steps in and says, 'Let's get him back into the yard.'

She leads Foxley off, limping, me following; the rest make a caravan as excited as the Children's Crusade. In the stable, with the single bulb lit above the stall, we can see that the swelling is the size of a tennis ball.

My brain starts to work again. I haven't read all those James Herriot books for nothing. 'Ice!' I shout, and run towards the house, across the back of the yard (where all adult eyes are still on Jack) to the mud room, where Auntie Liz keeps the freezer.

Contrary to belief, farming families do not eat fresh wholesome food; your average farmer and farmhand will eat more Twix, crisps and 'boughten' pork pies in a day than a *Clockwork Orange* urban droog in a week. With the simultaneous invention of the deep freeze/convenience food/the Meadow Market cash-and-carry, by the mid-seventies farmers were living on a diet of junk.

I open up the chest lid: above the mass of Findus Crispy Pancakes and Birds Eye Arctic Rolls (and, for form's sake, some stewed apple in plastic bags, with white metal ties) there are aluminium ice-cube trays. We run these under the tap, bash the cubes into the bucket. With wooden spatulas we shave off the thick ice fur on the freezer sides (the technology is in its infancy) and with reddening hands scoop this into the bucket.

We are now 'On a Mission', I tell my younger cousins in an attempt to make work fun. Fortunately Alys remembers we need something to pack the ice in, sneaks into the kitchen and takes two tea towels off the Aga rail (one, I recollect, was printed with scenes of Pembrokeshire) while Auntie Liz is on the phone.

Back to the stable. For an hour we pack ice on Foxley's bump.

Miracle of miracles, the swelling goes down. We

walk him round the yard. He seems fine.

Then a small puffing grey cloud of possible disaster comes driving up in the Land Rover. Uncle Ron has two stable girls, Mandy and Jane, who have been installed in a caravan at the back of the stables. (My cousin Olly, who, at twenty-one, is much the oldest of us, has smiled like a Cheshire tom cat with double cream on intravenous since they arrived.) Mandy and Jane were meant to be in Leominster buying some tack; they have forgotten the farm chequebook and have come back for it.

Jane-who-looks-good-in-jodhpurs is no fool. Why have we got Foxley out? 'Just exercising him.' She is not persuaded. So she takes him off us and walks him around. Then starts running her hands over and down his legs. I hold my breath.

Foxley passes muster.

'You were lucky,' hisses Mark, one of the tots.

Jack Reynolds, No. 6 fag drooping in corner of mouth and shirtsleeves rolled up, comes looking for us.

'Ment,' is all he says, rolling the cigarette from one side of his mouth to the other and back with his tongue.

The baler has been 'mended', is hitched to the Ford and we're off. Up the lane we go. Leysters Jack driving the DB with the front loader, Uncle Ron driving the Ford, and Olly at the wheel of the John Deere with the sledge behind it. (A sledge is a low flat trailer for carrying bales.) The rest of us are crammed into cabs, riding mudguards. I'm standing on the swaying rear hitch of the DB, one slip and I'm dead. But fun.

It's a convoy, good buddy. 'Come on and join our convoy . . .' Few things in life, even at fourteen when you can play polo, are as testosterone brilliant as being on a tractor on the way to get the job done.

The lane is narrow. Leysters Jack turns right into the Five Acre wheatfield, as does Ron. Olly misjudges the turn, and has to reverse the sledge for another go . . . The sound of the sledge cracking into the telegraph pole in the hedge is louder than three diesel engines combined.

We all stop, and de-tractor. The pole is broken in half like a matchstick, the top held up only by the wires. Then it starts to fall, and we run, and it comes crashing down, wires spinning like the arms of Shiva.

The top of the pole lands on the wooden bed of the sledge, thumping up a cloud of dust.

By some unseen command we all shuffle twenty yards up the lane, to the gateway of the field the telephone line crosses.

There are a dozen of us, and we gaze down the gentle valley at the course the telephone line takes. No one says anything for a minute or more, no one coughs or fidgets. It is late July, a yellowhammer is singing somewhere close by, and the green and gold fields stretch away until they melt into the distant heavenly shimmer. In the thirty years since, I have never quite put my finger on why the moment was so beatific. Perhaps because we were looking for something wrong, the cut line, and this only pointed up perfection.

Finally one of the Jacks: 'That's a hell of a view, that.'

We all agree it is.

But there is still the matter of the cut telephone line.

'Shouldn't we phone somebody about it?' asks Alistair, my cousin, Ron's other son.

Uncle Ron looks at him; his face flash-boils, and then he swipes Alistair across the back of the head. 'You bloody idiot!'

Everyone laughs except me. Because I was the bloody idiot who swung a metal crook into the leg of a hunter worth a thousand guineas.

IT IS NOT OFTEN in life – my life, at least – that actuality beats expectation. The Alvan reaper-binder surpasses my dreams.

I start down the centre of Flinders, to cut from the inside to the outside; if one cuts from the outside in, displaced animals congregate in the middle of the field and are slaughtered. So pronounced is the phenomenon that in agricultural lore the last sheaf used to be called 'the hare'.

The flat cutting bar is to the left of the tractor; this razes the wheat, which is seized by pincers and tied into a sheaf, which then pops along a short conveyor to drop behind the tractor. *Clickety-click. Clickety-click.* Machine music.

The Alvan reaper-binder never falters, never seizes. A perfect sheaf every time, all left in a row, making stacking them into a stook easier.

So much for the agricultural angle. For the field's wildlife, the reaper-binder is a destroyer, despite cutting in a conservation-friendly manner and cutting late in the season so that most ground-nesters have got their young away.

As the cutter proceeds, animals and birds hiding in the corn move to the side, as sea does before the bow of a ship. Mice and rats, for the most part, flee; one of the hares leaves it till the very last moment before scooting into the wildflower border, mazing as she goes, ears upright, and graceful, in the way that greyhounds and racehorses are graceful, as she ripples through the wheat.

Hares have large hearts to enable them to achieve such speed. Up to 1.8 per cent of body weight, compared to 0.3 per cent for a rabbit.

See how they run.

The red-legged partridge family slink away successfully, making the tiniest quakes in the standing wheat as they progress.

Rabbits do what rabbits do. Freeze. Sometimes I see them just in time and stop. Three catch the cutter and die.

The stone-squat toads, perfectly camouflaged and imperfectly slow, suffer most. For a fortnight or so they have been night walking (toads can only manage the feeblest hop) around Flinders, catching insects on their flicking tongues. They have come down from the wood on the pioneers' path, skirting the paddocks, in and out of the Chemical Brothers' field, through the hedge into Flinders. The fox comes the same way; yesterday, so did a snuffling hedgehog, to the bloody-nosed interest of the dogs.

By the time I've done half the field, I have carrion crows and buzzards shadowing me (a personal black cloud), much as birds followed me when I was ploughing. The crows and the buzzards are scavenging; the windhovers do not stoop to such unseemly depths today. Two patrol the field, dropping on the voles forced to flee their homes.

A cub fox is low in the edgeland grass, waiting for some heedless creature to scuttle to its trap-jaws.

Paradoxically, carnage is good. Carnage is good, and proof of biodiversity. While I'm cutting, the Chemical Brothers start on the twenty-acre wheatfield with their Lexion combine. No vultures circle there, no foxes crouch and wait. Why would they? There are no animals living there to be killed.

Flinders is meat on an earthenware plate for predators. They need to live too.

So do the seed-eaters. One crucial reason I wanted a reaper-binder is that, paradoxically, it is less efficient at scooping up every ear of grain, every seed of weed, than a combine. Some grain and seeds get knocked to the earth, making a meal for the smaller creatures, not only now but for months to come.

By the time I've done two thirds of the field, pink chaffinches, brown sparrows, black rooks, grey wood pigeons and yellow-and-red goldfinches are flying in for their share. Down they come through a perfume haze of thrashing cereal dust soil and pollen. I love the smell of a cut wheatfield in the afternoon.

Within three hours I've finished. Flinders is now a square expanse of pale nails on red earth, save for where the grass and weeds took their firmest hold, and there the wheaten spikes protrude from shaved green. Around the four edges, the wildflowers are still thick and luscious, especially the corn marigold, which splatters its gold generously.

With a boldness that teeters on insanity, the red-legs pick over the new land in broad daylight.

It's the same old beautiful field, just with a different face.

Before leaving, I look over the hedge at the Chemical Brothers' mute wheatfield. There is not a spilled grain to be seen, just chaff and strewn straw, as though a giant snail has left a yellow trail.

HERE IS THE WILDLIFE that used to be in the fields at Victorian harvest time, as recounted by Richard Jefferies:

All the village has been to the wheat-field with reaping-hooks, and wagons and horses, the whole strength of man has been employed upon it; little brown hands and large brown hands, blue eyes and dark eyes have been there searching about; all the intelligence of human beings has been brought to bear, and yet the stubble is not empty. Down there come again the ever-increasing clouds of sparrows; as a cloud rises here another cloud descends beyond it, a very mist and vapour as it were of wings. It makes one wonder to think where all the nests could have been; there could hardly have been enough caves and barns for all these to have been bred in. Every one of the multitude has a keen pair of eyes and a hungry beak, and every single individual finds something to eat in the stubble. Something that was not provided for them, crumbs that have escaped from this broad table, and there they are every day for weeks together, still finding food. If you will consider the incredible number of little mouths, and the busy rate at which they ply them hour by hour, you may imagine what an immense number of grains of wheat must have escaped man's hand, for you must remember that every time they peck they take a whole grain. Down, too, come the grey-blue wood pigeons and the wild turtle-doves. The singing linnets come in parties, the happy greenfinches, the streaked yellow-hammers, as if any one had delicately painted them in separate streaks, and not with a wash of colour, the brown buntings, chaffinches – out they come from the hazel copses, where the nuts are dropping, and the hedge berries turning red, and every one finds something to his liking. There are the seeds of the charlock and the thistle, and a hundred other little seeds, insects, and minute atom-like foods it needs a bird's eye to know. They are never still, they sweep up into the hedges and line the boughs, calling and talking, and away again to another rood of stubble without any order or plan of search, just sowing themselves about like wind-blown seeds. Up and down the day through with a zest never failing. It is beautiful to listen to them and watch them, if any one will stay under an oak by the nut-tree boughs, here the dragon-flies shoot to and fro in the shade as if the direct rays of the sun would burn their delicate wings; they hunt chiefly in the shade. The linnets will suddenly sweep up into the boughs and converse sweetly over your head.

THERE ARE 880 SHEAVES of wheat in Flinders. I know because I picked every one up to put them in 'stooks', of six sheaves leaning upright against each other in a tunnel. The sheaves need to be on end so rain will run off, but the sun and the wind catch them, and the 'weed' in the crop dies off.

'Back-breaking' would cover it, so would 'hand-hurting' since the sharp ends of the straw prick, as do thistles caught up in the sheaf. The MacWet gloves are a must. The sheaves have to be bumped down, so Jack Evans (who has taken up the role of supervisor) tells me, as though 'you're sticking spikes in the ground'. On three days when cloud and zephyr would please, the weather god sends three destined for the Sahara.

'All one's butterfly memories are sunny ones, bright pictures in the mind which colour the dark days of winter.'

BB (Denys Watkins-Pitchford)

There is a small white butterfly (*Pieris rapae*) flitty-floating in the border of Flinders, which now brings the species tally for the year to:

Small tortoiseshell	Small blue
Cabbage white	Meadow brown
Red admiral	Gatekeeper
Dark green fritillary	Small white
Peacock	

And in the Chemical Brothers' twenty-acre wheat-field? Gatekeeper and cabbage white.

It would be too much to claim that butterfly-philia is an absolutely British condition; the Russo-American Vladimir Nabokov – yes, the *Lolita* man – was an ardent entomologist, briefly the official curator of Lepidoptera at the Harvard Museum of Comparative Zoology. That said, the British, like mad dogs, seem genetically engineered for dashing about in the summer sun, the butterfly time. Lashings of tweed and eccentricity are virtual sine qua non for entomology. When the butterfly collector Lady Glanville died in 1709, her overlooked kin even tried to have her will set aside because 'none but those who were deprived of their

234

Senses would go in Pursuit of Butterflies'.

Probably the first purposeful collector of butterflies was the parson-naturalist John Ray, who was sticking butterflies to boards in the 1690s. Within forty years, there were sufficient like minds to form the Society of Aurelians, the first entomological club in the world. Being British it met in a pub, the Swan Tavern in London's Cornhill. Aurelians? With butterfly study in its infancy, these Georgian naturalists made up the etymology of entomology as they went along. Not all of their nomenclatural inventions were successful. Aurelian, meaning 'golden one', died on the wing.

The Victorians were big on collecting, and none was bigger than 22-stone, six-foot-three Walter Rothschild. The eccentric's eccentric, Lord Rothschild assembled a collection of 2.25 million set butterflies and moths ('no duplicates'). One of Rothschild's collectors, Albert Meek, earned enough from commissions to buy a cattle ranch in Queensland.

No one breaks a butterfly on a wheel, and in twenty-first-century Britain no one collects butterflies. In the 1970s, when Britain was loud with butterflies, my friend Tim and I would set off with fishing nets to catch them. I never pinned specimens to boards (too cack-handed; also I could not kill the bugs with chloroform or cyanide of potassium as recommended by the *Observer Book of Butterflies* because I had just watched the *World at War* episode on the Holocaust, which rather turned me against gassing things). But it's over. The butterfly collector has fallen from grace. Quite aside from the de-naturing of contemporary childhood, more about collecting apps for Apple iPhones than scrumping apples, there was always the terrible destructivity of butterfly collecting. Due to the collectors the ineffability of evolution is better understood; minute wing variation in the pretty marsh fritillary, discovered Professor E. B. Ford, was no accident but a real, living adjustment to environment. Neither can the collectors be blamed for the butterflies we have lost, in the Lady Bracknell understanding of the word, such as the English large copper, the chequered skipper, the English large blue . . . Pollution, agribusiness, infrastructural development are the true butterfly-killers.

What are butterflies for? The life of a butterfly is nectar-supping, flashy and short. Butterflies are only third-rate pollinators. In their various developmental stages they make convenient meals for predators, but in Britain no fauna depends on the survival of butterflies. Medieval boys fastened fine flaxen thread to the bodies of butterflies and flew them like tiny living kites. We frequently freight Lepidoptera with psychological meaning. The Open University once ornamented its brochures with butterfly images: yes, you too can metamorphose into a mature, scintillating creature.

As I write this in my be-flowered cornfield, a marbled white has flickered through, immediately enlivening the scene with soul and beauty. The design on a marbled white's wings is the full moon seen through the tracery of bare winter branches. Surely old John Ray was correct when he postulated that the ultimate purpose of butterflies was 'to adorn the world and delight the eyes of man'.

THE STOOKS ARE card houses. A vicious overnight wind has knocked half of them down. In the morning I try to reset them, but the wind whimsically fells the stooks as fast as I stand them up.

Sheaves prone on the ground will not dry. A magpie flickers across the field, making its death-rattle call.

Face rinsed with sweat, I collapse beneath the hedge and its pool of heavy shade. The scarlet pimpernel's drops of blood match the stalk-scratches on my hands.

There is a clump of shepherd's purse within reach; the seed pods, held to the light, do seem to be miniature coins in a clutch-bag. Shepherd's purse also goes by the folk name of mother's heart, since the seed pod is as heart-shaped as it is bag-like. Country children used to try and pull the pods from the stalk without the case splitting. If, as usually happened, the seeds burst out, the children would shout, 'You will one day break your mother's heart.'

29 AUGUST FLINDERS. Soft evening sky, caressed into pink cotton folds.

I've come to make some 'pop holes' in the fence between the two paddocks so that the hares, which now number eight, will have the run of three fields. Two hares, Flinders jill's leverets, are immobile in stubble. She herself lollops out from the stubble to pick at a sow thistle in the margin. Hares are ungainly unless running.

The Flinders hares are neither together nor apart in their society.

The fox has ignored them in favour of the rabbit colony which has started under the hedge with the Chemical Brothers.

As Jack points out, twenty years ago my hares would have been poached by now. Nobody poaches any more, save for gangs out from Birmingham after deer. One of the unintended consequences of cheap food is that it made poaching unworthy of the effort and time.

Jack gets as much enjoyment from watching the hares as I do; Flinders is the terminus of his constitutional. 'After you,' he says, 'there's not much to see, and those bloody tractors come whizzing round the bend almost knocking you over.'

But hare-watching is a pleasure, even a solace.

The Georgian poet and hymnodist William Cowper kept hares as pets. They were his cure for manic depression. One hare, Puss, lived until it was eleven years and eleven months old, dying 'of mere old age'.

Puss grew presently familiar, would leap into my lap, raise himself upon his hinder feet, and bite the hair from my temples. He would suffer me to take him up, and to carry him about in my arms, and has more than once fallen fast asleep upon my knee. He was ill three days, during which time I nursed him, kept him apart from his fellows, that they might not molest him, (for, like many other wild animals, they persecute one of their own species that is sick), and by constant care, and trying him with a variety of herbs, restored him to perfect health. No creature could be more grateful than my patient after his recovery; a sentiment which he most significantly expressed by licking my hand, first the back of it, then the palm, then every finger separately, then between all the fingers, as if anxious to leave no part of it unsaluted; a ceremony which he never performed but once again upon a similar occasion. Finding him extremely tractable, I made it my custom to carry him always after breakfast into the garden, where he hid himself generally under the leaves of a cucumber vine, sleeping or chewing the cud till evening; in the leaves also of that vine he found a favourite repast. I had not long habituated him to this taste of liberty, before he began to be impatient for the return of the time when he might enjoy it. He would invite me to the garden by drumming upon my knee and by a look of such expression, as it was not possible to misinterpret. If this rhetoric did not immediately succeed, he would take the skirt of my coat between his teeth, and pull it with all his force. Thus Puss might be said to be perfectly tamed, the shyness of his nature was done away, and on the whole it was visible by many symptoms, which I have not room to enumerate, that he was happier in human society than when shut up with his natural companions.

IN THE WILDFLOWER border a cornflower bends under the weight of a fat bumblebee. Listen to the bees. A bee about its business is never sad. 'The busy bee has no time for sorrow,' wrote William Blake. On the contrary, their susurration is contentment, akin to the purring of the cat.

An orange spider climbing a wheat stalk that escaped the chop tries to trap the last sunlight in its web.

The evening sun is glowing with pleasure on my old-time scene of stubble 'n' sheaves.

Even the rooks sitting on the sheaves, pulling at the ears of wheat, fail to spoil the scene's benedictions. Quite.

Death in the stubble I
'Harvest' is taken from the Old English *haerfest*, meaning 'autumn', and a sense of repose, of finished business, is tangible in the air.

That is my interpretation: the long-legged harvestman is *beginning* his killing business in the field.

Although the harvestman looks like a spider, he is not a spider. Peer closer: he does not possess a body divided into two distinct parts by a waist (as does the spider), nor does he dispatch his prey with poison fangs, or spin a web. Instead, the harvestman hunts down insect quarry on the stiltiest of legs and seizes it with pincer-palps – as this 5mm brown blob of a harvestman is doing to a small white day moth.

240

The moth tries to flap away, but the harvestman is on its back, squeezing it to death. When the eating time comes, the harvestman's mandibles are metronomic pistons. There is no sense of savouring, of pleasure.

Most species of harvestmen pass the winter as eggs and do not mature until late summer, harvest time.

Survival in the stubble
Hiding in the wildflower hem of Flinders is a portly patriarch of a toad. Toads can live in the wild for twenty years and adjust the hue of their warty skin to match local soil. So the back of this king toad is muddy-brown to match the clay on which he squats.

He waits, and waits. Then a harvestman – perhaps the moth-killer himself – scuttles through the wheat stalks along the edge of the stubble . . . The toad flicks out its tongue, which catches a leg of the harvestman. The harvestman, however, has a miraculous survival trick up its sleeve.

The harvestman has detachable limbs. The trapped leg is jettisoned, the harvestman runs on and lives another day.

Death in the stubble II
Amid the wildflowers, despite daylight and sunshine, it is moist and shady enough for a slug to slither about.

The snail is saved from our loathing by its house; the *sluggen* is all slime (a glandular secretion which prevents it drying out and helps it slip down the human throat; in the Middle Ages slugs were eaten alive as a TB cure, and were swallowed in milk to eat away ulcers). The smell organs of a slug, positioned on the creature's tentacles, are capable of detecting prey at a distance of several feet.

On the scent of something, the black slug elongates its shell-less body and forces its head into a soil cranny. And drags out a thin red worm.

These three incidents happened in the space of ten minutes while I was eating my lunch, and within touching distance.

The triple *kah* of the carrion crow in flight echoes off the harvested field.

CONFESSION: I WENT poaching as a teenager in Herefordshire in the 1980s, and I know the thrill of it. The tracking of game on someone else's land, the satisfaction of the kill, the jump at every *kerwick* of the owl on the way out.

I was caught putting night lines for trout into the River Lugg. Mind you, everyone in the rural west of England was poaching something then. Miss Giles from Box Tree Cottage, the upper-class spinster down on her uppers, used her green Austin Maxi for purposes unimagined by the manufacturer. She would shoo pheasants on to the lane, then climb in her car and speed towards them, bright-eyed with joy. Tom the milkman had a Certus folding .410 shotgun, the poacher's classic weapon, which he kept under the front seat of his van. If he saw anything worth putting in the pot, he used to wind down the front window and shoot, a rural version of a drive-by shooting.

My fine for contravening the eighth commandment was to buy the gillie two gills of whisky at the Crown and Anchor. Like the gamekeeper and the farmer, Mr Paynter took a generally benign view of dilettantes.

There was one professional gentleman of the night in our village, the happily alliterative Percy the Poach. He would steal anything; newcomers would find the trees in their garden chopped down, and keeping Percy warm in his cottage. Caught drunk in charge of a horse and trap, sly Perce was acquitted on a technicality because, as he gleefully informed the magistrate, Dolly was a pony.

1 SEPTEMBER FLINDERS. I do a hay-cut on the two paddocks, and though I leave a wide grass margin, the paddock hares move into Flinders, the first major movement of the hares for three months.

Flinders is now well stocked with hares, since Flinders jill has had two more leverets. There are ten hares hidden in the three-metre margin of the field, adjoining the paddock. Each form has its back to the wire stock fence. The hares are no fools; they realize that the fox is unlikely to be able to penetrate the fence; danger lies to the front, across the open field, the bed of straw nails.

THREE WEEKS AFTER harvest, I go to meet Rob Pryor at Flinders, since I've arranged for him to load and transport the sheaves in his silage trailers. He thinks he'll need only two trailers, I tell him he'll need three.

He arrives, looks at the stooks, says, 'I didn't think you'd get so much off. You're right. We'll need three trailers.' He is also arranging for two of his teenage nephews to help load the stooks (another job barely covered by 'spine-snapping').

We decide to haul on Monday.

On Sunday I fall off a horse. There are only two good ways to fall off a horse, be very, very young or very, very drunk, and I was very, very neither. I injure my hand. I can just about drive; I cannot pick up sheaves.

To note the beauty of the day,
And golden fields of corn survey;
Admire the pretty flowers
With their sweet smell;
To celebrate their Maker, and to tell

The marks of His great powers.

Thomas Traherne (c.1636–74), *Centuries of Meditations*

In my enforced rest I go to Hereford Cathedral on an overdue visit to look at the new stained-glass window installed to commemorate the seventeenth-century theologian and poet Thomas Traherne. He was local local, born in the city. Looking at the window, and the aphorisms selected in the accompanying leaflet, I am struck by the pantheism. This is landscape as the body of God: 'How do we know, but the world is that body; which the Deity has assumed to Manifest His Beauty' (*Centuries of Meditations*).

Traherne had a sense of the revelationary in everything he encountered, and believed men who ignored the beauty of nature to be 'statues dead'. Whereas those who gathered the beauty of nature were properly fulfilled:

> *To fly abroad like active bees,*
> *Among the hedges and the trees,*
> *To cull the dew that lies*
> *On every blade,*
> *From every blossom; till we lade*
> *Our minds, as they their thighs.*

*

TODAY I KILLED Bambi. Our friend Annie farms deer, and asked me to check on them while she was away.

245

When I arrived a fawn was lying on its front, in bracken, its back broken. Pitifully, the doe-eyed being tried to drag itself by the front legs, but its rear was an immovable anchor. The only mercy I could bring it was a 12-bore blast to the head. The sound of the gunshot pinballed around the valley, and disturbed the pheasants on the hill above me so they *cokk*ed manically, and cordite choked the air . . .

I stop at Flinders on the way home, as evening is falling.

A young kestrel, looking like its streaky brown mother, tracks over the field, then flies to perch on the telephone wires. A collared dove sits companionably beside it.

Suddenly, the kestrel plummets down, starts hovering, and a shape rises to meet it, and flails with paws.

The kestrel lifts, and speeds away, ashamed.

A hare has beaten a hawk.

In the wood, a wood pigeon, a notoriously late nester, is sitting squat on china-white eggs in a flimsy nest in the dark beech. Her mate is singing to her. The *coo-coo*ing comes fondly and forgiving across the stubble and the stooks.

NOT UNTIL 28 SEPTEMBER do we transport the sheaves from Flinders. It's the eve of the fat harvest moon, so we are able to work after the 'official' sunset. There is light from the sinking sun, light from the rising moon, and

246

we don't leave till after 9pm. Rob and his nephews go first, I follow. Convoy.

Unloading is simple, we just tip up the trailers and let the sheaves roll out on to the yard.

We have pigs, ergo we have rats, who will be delighted by the meal delivered by so many wheels. So I put the terriers' kennels next to the stack.

Sorted.

*

A YEAR IN THE LIFE of a conventionally farmed arable field: maize is the duck-billed platypus of plants, whimsically assembled from bits of others. The roots are from mangrove, the tall stems from bamboo. The hairy tops of cobs are beyond flora; they are joke-shop wigs. When the wind blows, maize is a raving crowd. Get your hands in the air!

Maize is cut in October–November when the starch levels are up.

21 OCTOBER In comes the forage harvester, which cuts in huge swathes. The chopped maize pours out, grey-green, from the pipe into the trailer of the tractor.

25 OCTOBER The field is harrowed.

27 OCTOBER The field is drilled with winter wheat.

6 NOVEMBER The field is already hinting green.

There is no rest for the land.

247

31 AUGUST FLINDERS. The dried poppy heads, reminiscent of urns, rattle like a child's toy. When the wind blows, the poppy heads rock and the fifty thousand pepper-black seeds are shaken out of the pot.

A blackbird sings plangent across the mist on the stubble.

There's a peal of bells from over the hill calling the faithful to Harvest Festival. The church spire is reproachful in the mist.

RUST NEVER SLEEPS, farming never stops. Feed the livestock, clean out the livestock, bed down the livestock. They eat, they defecate, I slave. Repeat.

So: it's late on a tyre-black night, and I'm on the way back from dropping off some joints of frozen lamb to a customer in Ross, when I finally get to check the sheep in the paddocks next to Flinders.

A rising wind alternately swaddles and strips me.

The sheeps' eyes are twin emeralds in the beam of the 1.2-million-candle-power torch. All, I thank the Lord, is well. No sheep seem to be missing. Or dying. Because a sheep's most earnest wish in life is death.

As I turn for the Land Rover, there is a pause in the wind, and I hear a dog barking in the wood.

No, not a fox: a domestic dog, of a yappy sort.

There was a stray dog in a field up by Pool Farm last week, running the sheep there ragged.

Into the storm-flounced wood I go in search of the

miscreant dog, the torch search-lighting the arboreal pillars, living and upright, fallen and dead. There is no dog to be seen.

In the moment when I am standing in the sucking dingle mire, where the kingcups flower in spring, in the precise centre of the wood, the bulb of the torch blows . . . I'm in a blackout. A black hole. I am stranded in the chaos that came before Creation.

I cannot see a single thing. I'm not scared; I like the dark. I would, however, quite like to get out of the wood without injury and ASAP. Apprehensive, then.

Over the summer, I spent a moment or two listening to the wind soughing in the trees of the wood. (In books, wind always soughs.) Under more extreme circumstances than I had intended, my self-imprisonment in the wood is an examination of my knowledge.

I remember: wind in larch is a fine-sieved jet-roar whoosh. Wind in oak is the shimmying of aluminium foil. Wind in beech tops is pattering silver-money music.

I know that if I head uphill, with the larch to my left, and find the place where the oaks greet the beech, I can get on to the ride, the path, which bisects the wood, and then out.

It's about fifty yards to the ride, a minute's walk with eyes and without obstacles.

I inch forward, larch sound to my left, with a stick in my left hand, blind-man tapping, the torch in a clenched-fist Marxist salute in my right hand: this to stop branches whipping my face.

Trees creak and crack. The entire world is movement and sound.

After five minutes or so of stumbling, I detect the sheet rattle of oaks. There is a further guide to navigation: the spicy smell of antique oak-leaf moulder, blown towards me by the wind, hits the septum like snuff.

I'm out on the ride, and can see the cottage lights of Orcop Hill to steer by.

As I walk across the paddocks into Flinders I am aware of relief and relative tranquillity. It occurs to me that when the Neolithic farmer conquered the wildwood he must have felt the same peace and pleasure in cultivated, controlled land. Out, at last, of the wildwoods, and into the fields.

CHAPTER VI

The Turn of the Earth

*From the sowing to the reaping, the wheat-field
gives a constant dole like the monasteries of
old, only here it is no crust, but a free and
bountiful largess.*

Richard Jefferies, 'Walks in the
Wheat-fields'

THE THRESHING FLAIL IS an ancient instrument to
part the grain from the husk. It consists of two
pieces of wood, the helve and the beater, joined by a
thong. The helve is a light rod several feet long, the
beater a shorter piece. With a flail, one man can thresh
seven bushels of wheat, eight of rye, fifteen of barley,
eighteen of oats a day (one bushel equals about 35 litres).
The flail remained the principal method of threshing
until the mid-nineteenth century, when mechanical
threshers became widespread.

I thought I would have a go at hand threshing, to get
grain for some homemade bread.

Although making a flail is easy enough, I buy one.

As I say, you can purchase anything on eBay, including a wooden wheat flail from the US for $25, shipped that day ('Rare antique wood wheat flail primitive grain threshing farm tool rustic décor').

Cometh the threshing hour: I lay down a large tarpaulin on the stone floor of the barn, lay down several sheaves of wheat, and batter away. The trick, I soon discover, is to hold the flail handle almost horizontally to the side, so that the 'swipple' (the head) is spinning like a Catherine wheel. If you stand and hit vertically down, the flail hits the user painfully in the face when it is brought back up.

Take it from one who knows: there is a reason the agricultural flail was transmogrified into a weapon of war in the Middle Ages.

Hand threshing works. Period. After I've battered away, the threshed wheat is tossed into the air ('winnowing'); the breeze takes away the lighter chaff, the grains fall to the ground. Collected, I have three bags full, and in a sort of exhilaration dive my arm deep into one of the sacks, so that the grain runs up to my elbow. Then I grab a handful, and let the grain fall like a golden stream back into the sack.

Like many a job in old-style agriculture, threshing is hard graft. The mechanization of agriculture freed men from such drudgery, but really only to become slaves to machines in factories and mills.

On the land, threshing was sought after, because it was employment in the dry, in winter.

MY INTENTION NEVER was wheat for bread; my intention was wheat as food for our cattle, sheep, pigs and chickens. To thresh and mill the wheat for livestock is a five-minute job a day; I put the day's allotment of sheaves on the barn floor, and run the tractor and field-roller over them. Cereal is easier for livestock to digest if the grains have been broken open.

The roller weighs about 620kg. I then pick up the wheat stalks and chuck them in the cattle feeder; the chickens, pigs and sheep get the rest, either neat or mixed up with concentrate as the case may be.

Sometimes, to keep the chickens entertained, I hang sheaves upside down in their paddocks, so they can pluck whole grains off.

By a rough calculation, Flinders gives me 5 tons of grain (roughly 1 ton per acre). I could have improved the tonnage per acre by a more sensible drilling policy, and with better 'weed' control, without too much detriment to my wildlife policy. The real benefit for conservation has been the wildflowers at the edge of the ploughland, and the grain and seeds left behind from the harvest for gleaning by birds and animals.

One of the great myths of agriculture is that conventional farming is, by a country mile, more productive than organic farming. On the contrary: as *Nature* magazine reported in 2000, following one of the biggest agricultural experiments ever conducted, Chinese scientists tested the key principles of modern rice-growing (planting a single hi-tech variety) against a much older technique (planting several kinds in one field); farmers who reverted to the old method enjoyed an increase in yield, and a decrease in 'rice blast' (a fungus) of 94 per cent. The farmers planting a mixture of strains were able to stop applying their poisons altogether, while producing 18 per cent more rice.

Nature also reported in 1998 that yields of organic maize are identical to maize grown with fertilizers and pesticides, while soil quality in organic fields dramatically improves. In trials in Hertfordshire, wheat grown with manure has produced higher yields for the past 150 years than wheat grown with artificial nutrients.

By growing and milling my own non-GM, non-chemical wheat I saved £30 a ton in livestock food.

And what price for all the animals, birds and bees that Flinders succoured?

15 SEPTEMBER I'M transporting a small flock of sheep to graze the paddocks at Flinders.

The starlings drift in yobby aimless bands, teenagers in a shopping centre. They fly up to the electricity wires, sing, then decide it is not high enough and try the alders instead.

A hornet cruises past; it's the Provisional wing of the wasp family.

Chaffinches are feeding quietly at the edge of the

field; they appear to nod their heads as they walk. Rather they keep their heads still, eyes focused for food, as the body moves forward, then move their heads to catch up with the body. The chaffinches will not stray far from Flinders this winter.

21 SEPTEMBER Swallows congregate on the barn roofs at Pool Farm. Out on the fields around Flinders, the ploughs and drills are busy once again, sowing winter wheat and barley.

No, there is no rest for the land.

Jackdaws play in the sky, then drop like lead weights.

1 OCTOBER Three degrees of frost this morning; yet by lunchtime it is warm and I am in shirtsleeves.

A good autumn morning should be like a good wine: a complex bouquet, with notes of liquorice from falling, rotting leaves; the tart acid of hedge fruits; the musk of rams; the itchy pickle of silage.

There are also three yellowhammers in Flinders, a flash dab of colour among the brown sparrows and the pale stubble wreckage.

Daddy-long-legs drift into Flinders from the paddock; one daddy-long-legs gets caught in a clump of shepherd's purse.

The wasps attack the daddy-long-legs, rip its gauzy wings off and one conveys it away.

The daddy-long-legs will be a meal for the million wasp young.

Goldcrests are a little cloud of noise in the hedge; their *tsee, tsee* is the squeak of a jewellery box hinge. Three hen pheasants scuttle in, so small I think they are partridges.

'TIS A FAMILIAR refrain: 'Oh, how lovely it must be to combine farming and writing. Hand and head. Holistic.'

They don't combine. By the law of Sod, every writing deadline/engagement will synchronize with an animal emergency. The nadir: the *Times* Literary Festival; our sheep escape from rented land; Penny has to pick me up; I arrive in Cheltenham in my boiler suit and wellingtons, having forgotten to ask her to bring my writer's uniform of blue jacket and jeans. Surreal: I stand, in said boiler suit and wellies, next to Henry Winkler (The Fonz!) in the queue in the greenroom for tea. Aayyy!

Someone at the festival, quite understandably mistaking me for one of the maintenance men, asks if I can take a look at a leaking pipe in the male loo; one of the fruity young girls with a clipboard who intern at such things asks me to move some cable. At this rate of employment I may swap careers to become a DIY man.

So: on the day I'm meant to be giving a talk in Knutsford, a sow decides to give birth early. Having played midwife, it's pedal to metal to Cheshire. I arrive in Knutsford to find that the corner shop sells McLaren cars. I'm in WAG country. But lovely people at the literature festival. I am asked the inevitable question: 'How does one start up in nature writing?' I give the only sensible answer: 'Go outside, and stay there.'

Afterwards, the chair, Liz, introduces me to Ray Rush, a stooped, octogenarian retired farmer ('a bit of a local celebrity'). Liz insists I go and see Ray's corn dollies. I'm not expecting much, but follow them in my car to Siddington Church, where Ray has his corn dollies displayed.

The church is plain odd; a folly, a Victorian version of a black-and-white Tudor church. Ray's farm is next door. 'WELCOME' is spelled out in letters made from runner beans in the church porch, which at this harvest time is full of real fruit and veg; funny in shape, with dirt on. Proper harvest goods.

I open the door of the church – and it is the eighth wonder of the world. There are a thousand corn dollies made by Ray. They are everywhere, on every surface. A corn-dolly clock; crosses, crooks, flowers, geometric designs, people. All put up and taken down by Ray. Every year.

Ray takes me to his farm next door, to his workshop where he makes the dollies, which require a special straw, Maris Wigeon. His cats are pleased to be let in to chase mice. On the way back to my car, he points out the old cowshed: 'That's where we used to milk thirty Jerseys.'

Thirty years ago, a man or a woman could make a good living on thirty milk cows.

Liz has secretly bought me one of Ray's corn dollies, and gives it to me as a present as I get into the car. If you want faith restored, go to Siddington Church at Harvest Festival.

Driving home, I wonder: who makes corn dollies now? My grandmother did. Ray still does, but who else, outside of a craft 'workshop'?

A DISCOURSE ON corn dollies: 'dolly' is probably a corruption of 'idol', or the Greek word *eidolon*, meaning apparition.

Long, long before the coming of Christianity to Europe, it was believed that the spirit of corn lived among the crop, and she took refuge in the last sheaf to be harvested. To give the corn-spirit a refuge, these final straws were woven by the reapers into the likeness of a woman or into a geometrical cage. Then the effigy containing the corn-spirit would be carried home to the farmstead, and hung on the wall the winter long. In spring, the dolly would be taken out into the field, and ploughed into the first furrow of the season.

There are numerous variations on this basic corn-spirit rite. In some parts of Scotland the last corn to be cut was the 'Old Wife' (Cailleach), who, after being corn-dollied up, was given to the horses to eat when they went spring ploughing; in still other parts of Scotland, along with the north of England, reapers ritually sickled the last corn or 'kirn', blindfolded and with general merriment, and dressed it up as a child's doll. The kirn-baby was both a fertility and a good-luck charm. The reapers of north Pembrokeshire also regarded the cutting of the last standing corn as a ceremony as rich in hilarity as it was in significance; all harvesters present threw their sickles at it, and the one who succeeded in cutting it received a jug of home-brewed ale. A 'Hag' was then fashioned and taken to the farmhouse. In Devon and Cornwall a sheaf of ears was made into 'the neck', to become the subject of a raucous ritual known as 'crying the neck'. The folklorist William Hone was witness in 1836:

An old man, or someone well acquainted with the ceremonies used on the occasion (when the labourers are reaping the last field of wheat), goes round to the shooks and sheaves, and picks out a little bundle of all the best ears he can find; this bundle he ties up very neat and trim, and plaits and arranges the straws tastefully. This is called 'the neck' of wheat, or wheaten ears. After the field is cut out, and the pitcher (of cider) once more circulated, the reapers, binders and the women stand round in a circle. The person with 'the neck' stands in the centre grasping it with both hands. He first stoops and holds it near the ground and all the men forming the ring take off their hats, stooping and holding them with both hands towards the ground. They then all begin at once in a very prolonged and harmonious tone to cry 'the neck!' at the same time slowly raising themselves upright, and elevating their arms and hats above their heads; the person with 'the neck' also raising it on high. This is done three times. Then they change their cry to 'wee yen! wee yen!' which they sound in the same prolonged manner as before, with singular harmony and effect, three times. This last cry is accompanied by the same movements of the body and arms as in crying 'the neck.'

After having thus repeated 'the neck' three times, and 'wee yen' or 'way yen' as often, they all burst out into a kind of loud and joyous laugh, flinging their hats and caps into the air, capering about and perhaps kissing the girls. One of them gets 'the neck' and runs as hard as he can to the farm-house, where the dairy-maid, or one of the young female domestics, stands at the door prepared with a pail of water. If he who holds 'the neck' can manage to get into the house, in any way unseen, or openly, by any other way than the door at which the girl stands with the pail of water, then he may lawfully kiss her; but if otherwise, he is regularly soused with the contents of the bucket.

On a fine still autumn evening the 'crying of the neck' has a wonderful effect at a distance, far finer than that of the Turkish muezzin, which Lord Byron eulogizes so much, and which he says is preferable to all the bells in Christendom. I have once or twice heard upwards of twenty men cry it, and sometimes joined by an equal number of female voices. About three years back, on some high grounds, where our people were harvesting, I heard six or seven 'necks' cried in one night, although I know that some were four miles off. They are heard throughout the quiet evening air, at a considerable distance sometimes. But I think that the practice is beginning to decline of late, and many farmers and their men do not care about keeping up this old custom.

In *The Golden Bough* the anthropologist James George Frazer pointed out that corn-dolly rites, whatever variant one considers, are Stone Age primitive since 'there are no priests'. The rites are communal, to be performed by anyone, anywhere. Also, the ceremony is magical rather than propitiatory; the bountiful harvest is to be obtained by influencing nature rather than sacrificing to, or pleading with, the gods.

Mechanization in the nineteenth century began the eradication of the corn dolly, although vestiges of sympathetic magic lingered on stubbornly. In 1911, in Northumberland, when the last sheaf was taken reapers declared they had 'got the kern'; the sheaf, dressed in a white frock, was hoisted on a pole to preside over a harvest supper. 'Crying the neck' was observed as late as 1930 in Devon.

Modern harvesting eliminated the sheaves from which corn dollies would have been fashioned; you can't make a corn dolly out of wheat that has been mangled through a combine. Also modern cereals are too thick and too short for decorative straw craft.

I hang my corn dolly from Ray on the hall wall. But it is so lovely, I can't bear to consider breaking it open in spring as stimulus to the land.

16 OCTOBER Only the linking of its branches stops the lane-side hedge being swept away by the gale. Like protestors, linked arms facing riot police, it stands its ground.

Leaves whirl across Flinders.

On the next day it rains, and the worms take the leaves down into the earth to nourish it.

24 OCTOBER Redwings, those nocturnal wanderers, arrived in the night, on the north wind. If the swallow signals summer then the redwing heralds winter's arrival.

Ten redwings are trying to pick over Flinders but the wind is too strong for the small thrushes. They repeatedly protest *tsstick*. Nonetheless they are blown off into the wood.

The grey heavy hulls of wood pigeon remain moored to the Flinders surface.

The wind is blowing out the last of the wildflowers, as if they were candles. Only the flaming-red poppy is still alight.

26 OCTOBER Although the Flinders hedges have, in the past, been trimmed in the manner of suburban privet they have now fruited with haws, hips and sloes.

The redwings are plundering the hawthorn's berries, lipstick kisses in the blue sky.

A snapshot of the birds in Flinders in one second, one frame: rooks, house sparrows, chaffinches, red-legged partridge, and goldfinch young on thistles bowing with the weight of the passing year.

Flinders' resident chaffinches have been joined by a dozen others, from northern Britain or northern Europe, and I doubt it matters to the Flinders birds from where the immigrants hail. They are all competition.

27 OCTOBER Daylight is crushed away between the grindstones of Night and Earth.

I'm checking sheep; something white catches my eye as I walk across Flinders in the gloom of all the centuries. One of the partridges has been killed; its feathers lie in a round cushion in the corner by the stock fence.

If I were a nature detective, I would say a sparrowhawk is responsible.

All the arable fields surrounding Flinders are now re-tilled and re-sown. Flinders is alone in its stubbleness.

Some floppy, weary crows pass overhead, suitably silent for once.

FROST-SHINE ON Flinders, jackdaws playing their morning games in the sky: I am bemused by a speckling of dark, ice-free circles in the stubble.

Jack Evans appears at the lane gate, beckons me over, breath wheezy in the cold. There were fieldfares in the Flinders stubble last night, 'cooched down'. Roosting.

I wish I had seen them.

Then the air drifts with the silver-bell tinkle of goldfinches to charm the field and me.

IN THE STUBBLE of the wheatfield a cock pheasant wreathes in and out of the vapour. He is a joker playing hide-and-seek.

On this morning of chalk mist I am digging a hole for a new gatepost. I can see about thirty yards into the field, before the white shrouds cut me off. There is just me and the dog, the muffled medieval quiet, the churchy smell of rotting leaves in the hedge-bottom, the wildlife of the stubble.

Brume clings to the webs of money spiders strung between the shorn wheat stalks. The incy wincy spider waits beneath its web to tie its victim in silk bonds. Or the spider does so if not eaten first by the pied wagtails who have taken up residence, most of them Pierrot-faced youngsters. John Clare described the motion of the wag-tail as 'tittering, tottering'. 'Twerking' would also serve.

There is a stark, minimalist beauty to stubble. The sharpness of the spikes, the barrenness of the earth, are suited to winter in a way that autumn-sown wheat, which grows soft and green over Samhain's season, never is.

The red-legged partridges scurry past on an unknown errand. On the edge of vision a young rabbit lopes along. Edith the Labrador drools a long string of glisten-ing spittle from a corner of her mouth.

A wood mouse trundles from the hedge-bottom into the field to gather seeds from the wildflower border.

A robin sings melancholically (of course), defending its territory, the only interruption to the winter-scented silence except for the chip-chip of my gravedigger's spade.

If I am honest, though, being isolated by November mist is downright creepy. Small wonder, I think, that in the traditional calendar 2 November – today – was the time of All Souls, when the ghosts of the dead returned to their homes. The air is thick with expectation.

The spade cleaves through the old red clay. It is like carving wood; the sides of the hole are polished smooth by the passing of the steel blade. As the spade goes deeper, so the colour of the earth changes down through a colour chart of pink to Victorian strawberry fool.

My mind drifts. What else is it to do, when one is digging a hole? I muse on Novembers past. November is the month of Martinmas, when traditionally in Herefordshire beef was smoked in the farm chimney, and the pig was slaughtered.

Then the robin alights beside me and eats a worm. I am brought back to earth, and go across Flinders digging holes in a repeat of the worm count of January. There are more worms per hole, seven or so. A gain of two worms per hole.

> And grey beard jackdaws noising as they fly.
> The crowds of starnels whizz and hurry by,
> And darken like a clod the evening sky.
> The larks like thunder rise and suthy round,
> Then drop and nestle in the stubble ground.
>
> John Clare, *The Shepherd's Calendar*

3 NOVEMBER In the subfusc of early dusk I arrive at the field. Blackbirds make their November music.

I have watched an autumn roost of starlings here for three days, and the spectacle has grown greater by the day.

At first I can't see them, but then gangs of starlings, in their winter plumage, fly in from six directions at once, and commence their synchronized swirling in the sky.

A starling is a featherweight on its own, but in numbers can cause damage. On 12 August 1949, starlings roosting on the minute hand of Big Ben caused the clock to run slowly, to the consternation of listeners to BBC radio expecting to hear Big Ben's chimes. Probably 1967 saw peak starling numbers in the UK, with an estimated 37 million. The birds were drawn to the heat of city centres, and huge flocks gathered over Birmingham, Manchester, Leeds, Newcastle, Belfast, Liverpool, Edinburgh and Glasgow. Councils tried to deter them with loudspeakers and flashing lights. The burghers of Glasgow tried bagpipes.

Today, you are more likely to see the autumn flocks in rural areas, like here. Within minutes the flock has grown to a hundred, five hundred, a thousand, as small bands come from every direction. The murmuration grows all the time, as does my appreciation of that collective noun, which is precisely onomatopoeic. And very old; the first known usage of 'murmuration' is in *The Book of St Albans*, 1486.

The flock swoops down to twenty, thirty feet above my head, their wings whispering a secret never to be told to humans. They are always in sync, never colliding; guided by some invisible conductor; they are one creature, a smoke dragon. Do they perform the aerial patterns to confuse hawks, or just for the joy of it?

Murmurations have caused poets to ponder the right metaphor for their shape and prescience; John Clare came close with likening a mass of starnels in the sky to a 'clod'.

The starlings fall like leaves into the wood.

Starlings have decreased by 80 per cent in forty years. They are now on the critical list of birds most at risk. The causes of the decline are not understood fully, although the sensible mind points, like a compass needle finding north, to intensive farming methods and changes in architecture. New buildings lack the nooks starlings love. The antiquarian John Aubrey mused that the crannies in Stonehenge were stone nestboxes carved by the Druids for their sacred bird.

When I go into the wood next day the larches in which the starlings roosted are covered with white shite, an early snow. So choking dense is the smell of ammonia I cannot get within twenty feet without retching.

For two weeks the birds roost in the wood. Then they leave. Behind them are larches ailing and sickly from their excrement.

OBSERVATIONS ON THE autumn hedges of Flinders, which are fountaining with unofficial bramble: the ivy, the flower of fidelity, is in bloom. It is the subtlest of flowers, but just as there is pleasure in gentle music, so there is in the light-green tones of ivy blossom. The hips of the dog rose, meanwhile, are violent blood-pricks on the blue sky. Almost all the leaves of the hedge are gone; jaundiced hazel provides occasional cover. Fig leaves. I can see past the foliage screen of summer, into the hedge's inner life.

I disturb a hedgehog, barrel-fat and ready for hibernation, who quick-trots through the interior of the hedge to become lost in the ivy-covered ruins. The woodmouse emerges from its burrow to lazily climb the blackberry and seize the last remaining fruit. Back on the ground the mouse bounds like a kangaroo, as do the grey sparrows who live in the hedge. Due to their evil past, it is said, sparrows have their legs fastened by invisible bonds to prevent them walking. At Christ's crucifixion on Calvary the swallow flew off with the executioner's nails but the sparrow returned them, holding them up for the hammer so they could be driven into the Messiah's hands and feet. In punishment, the sparrow is condemned to hop. For ever and ever. Amen.

In the field margin under the laneside hedge, where the grass heads are blown and empty and wellington-topping high, the hen pheasant and her chick make a run for it. How quickly children grow up. The chick is a poult now. There is a 'good crop' of weeds under the hedge – hogweed, thistle, cow parsley, docks – attended by an equally good crop of goldfinches.

The ubiquitous four and twenty blackbirds? They boing on their legs; concentrate their hearing towards the ground; boing; concentrate; boing. Repeat. We all have our predestination, our teleology. The blackbirds keep tight to the hedge, rarely going more than ten yards into the field. Five redwings out in the centre of Flinders' stubble glow gold with the warmth of the day.

Hedges are English icons. Once, the mother of our Polish friend Basia visited us; she was agape, Basia explained in the embarrassed way you do when interpreting your parents, because of the hedges. In *Matka*'s neck of the European plain there were no hedges; indeed, trees were at a premium.

The first hedges in England were simply the bits of woodland left over from clearing land for farming, and were used as a boundary between neighbours. Much of the arable agriculture of medieval times was carried out in a relatively open landscape; only with the Enclosure Acts between 1720 and 1840 did the great hedging of England occur. (No, enclosure was not all bad.) Under enclosure legislation, landowners could claim common land by partitioning it with walls or hedges; about 200,000 miles of hedges were duly planted, with hawthorn and blackthorn being the standard constituents.

A carrion crow, glimpsed over the hedge, is eating the eyes of the car-killed badger on the lane. By eating eyes, the carrion crow is able to see into the future. The crow lands in the field and wipes his bill, glistening with vitreous humour, on the napkin of the soil.

It starts drizzling. Or mizzling. Eskimos have fifty words for snow; Herefordians have fifty words for precipitation.

I am patrolling the hedges forensically, in the style of an East German border guard, for signs of the fox family sneaking into Flinders under the stock fence. There is no wind, and the taint of fox hangs in the dampening air.

Three magpies are recceing the furzy blackthorn, one hopping on top, two walking alongside. This is a clever if fruitless cooperation: the young of the late-nesting blackbird flew a week ago. The bowl-nest is bare, already filling with the detritus of hedge life.

The avian teamwork does not last for long. Magpies are singular, excitable birds. Watch them in the examining sunlight: they split three ways. One stays in the field; he walks left, runs right, zig-zags, tires of his erratic pedestrianism, then flies for thirty yards, lands, starts over. He finds a grub, strikes with his broad sword bill. Then flies off, low and undulating, as though following the contours of waves on an invisible sea.

The old cock chaffinch by the gate is still going, in the pink of health, despite time passing and the invasion of migrants. We have an understanding. I've brought some heads of wheat from home especially for him; all I ask is that he does not try to sing. He disobeys, and *chinks* away happily. Chaffinches have regional accents, and I am sure he has the local, 'erefordian, dropped 'h' high-pitch.

Night: THE GREASE-GLEAM of moonlight on the soil's skin; the same light glitters romantically on water, even ditch water. Stars are ice grains in the black sky. The cold is true purifying cold. Sabre-tooth-tiger-time cold. The *kerwicks* of the owl skip over the ground as flat pebbles skim water.

Weird. Across in the paddock, the horses, despite plentiful grass, manically attack the elm in the hedge, stripping the branches and trunks of bark so the elms become a white boneyard in the dark.

Richard Jefferies, on the Victorian field hedge:

On the other side the plough has left a narrow strip of green running along the hedge: the horses, requiring some space in which to turn at the end of each furrow, could not draw the share any nearer, and on this narrow strip the weeds and wild flowers flourish. The light-sulphur-coloured charlock is scattered everywhere – out among the corn, too, for no cleaning seems capable of eradicating this plant; the seeds will linger in the earth and retain their germinating power for a length of time, till the plough brings them near enough to the surface, when they are sure to shoot up unless the pigeons find them. Here also may be found the wild garlic, which sometimes gets among the wheat and lends an onion-like flavour to the bread. It grows, too, on the edge of the low chalky banks overhanging the narrow waggon track, whose ruts are deep in the rubble – worn so in winter.

Such places, close to cultivated land yet undisturbed, are the best in which to look for wild flowers; and on the narrow strip beside the hedge and on the crumbling rubble bank of the rough track may be found a greater variety than by searching the broad acres beyond. In the season the large white bell-like flowers of the convolvulus will climb over the hawthorn, and the lesser striped kind will creep along the ground. The pink pimpernel hides on the very verge of the corn, which presently will be strewn with the beautiful 'bluebottle', than whose exquisite hue there is nothing more lovely in our fields. The great scarlet poppy with the black centre, and 'eggs and butter' – curious name for a flower – will, of course, be there: the latter often flourishes on a high elevation, on the very ridges, provided only the plough has been near.

All silk and flame: a scarlet cup, perfect edged all around, seen among the wild grass far away, like a burning coal fallen from Heaven's altars.

John Ruskin, *Proserpina: Studies of Wayside Flowers*, 1879–86

THE LAST POPPY in Flinders has died today, the 11th of the 11th.

The corn poppy's relationship with humans is absolute and long. It inhabits the places where humans turn soil, dig graves or, as in the case of the Western Front 1914–18, shell the earth. *Papaver rhoeas* appears in late spring, blooms in early summer with a four-petal head; the heavy seed pods weigh down the flower, before the elongated oval pod explodes, casting thousands of black seeds to the wind, guaranteeing more poppies. The plant then dies.

The poppy was the plant of death and remembrance long before the Great War. A site in southern Spain, the Cave of Bats, yielded in 1935 fossilized poppy capsules which suggested that poppies were placed in Neanderthal graves c.4000 BC. The capsules, along with locks of hair, were tucked inside woven grass baskets, laid among human skeletons. The poppy featured in Greek funerary rituals and Persephone wore the poppy as a symbol of her deathlike status when imprisoned in the Underworld. The annual ritual of the Eleusinian Mysteries held just outside Athens had its origins in an agrarian cult which stretched back to Mycenaean times. Thousands of citizens made the pilgrimage; as they approached the town the initiates were scrutinized by two giant statues of Demeter: she wore on her head a basket decorated with the poppy and symbols of the harvest, and all around the temple were carved stone images of the flowers. Until the conquest by chemical farming in the 1970s, where you had corn you had poppies.

Do as I'm doing now and break the poppy's stem, and a milky sap oozes out, white blood. The sap contains the alkaloid rhoeadine, a sedative, which has been used in medicine since ancient times. Dioscorides, the physician who accompanied Roman armies and wrote the magisterial *De Materia Medica* between AD 50 and 70, said the corn poppy was a medicine: after boiling five or six little heads with three cups of wine and reducing the liquid by heat to half, 'give it to drink to those whom you would make sleep'. The white sap of the corn poppy is only mildly narcotic, but the Romantic poets infused the corn poppy with the mind-altering qualities of opium. In 'To Autumn', John Keats wedded the corn poppy with the opium poppy's timeless tranquillity: 'Or on a half-reap'd furrow sound asleep, / Drowned with the fume of poppies'.

Art nouveau, which flourished from 1890 to 1910, was a belated physical form of Romantic poetry. The motifs of curving flowers spoke to nostalgia for the innocent, pre-industrial age. An escape from dehumanizing industrialization.

The fields of Flanders were heavy with irony. The poppy grew well on the Western Front because of the fertilizer that was the blood and bone of men and horses, and the nitrogen from high explosives. Before the war, much of the soil of Belgium was too poor for an abundance of poppies. Captain J. C. Dunn, a medical officer attached to the Royal Welsh Fusiliers and author of the memoir *The War the Infantry Knew*, noted on the eve of the Battle of the Somme that the fields of his native Norfolk were far richer in poppies. Corpses and bombs were the making of the Flanders poppy.

On the 11th of the 11th I remember the dead, and the lost land.

AT HOME, FEEDING the cattle with the grain and the straw from Flinders.

The warm yellow of the straw is reassuring on a winter's day under a dismal sky; the golden grains are atoms of converted sunshine.

19 NOVEMBER Flinders. The consuming fires of autumn have almost finished devouring the leaves of the hedges.

There must be fifty wood pigeons on the floor of the field. They rise with their familiar linen-flap noise, but too slowly. A sparrowhawk takes one mid-air; the impact of the sparrowhawk's talons causes pigeon feathers to become dislodged and float reluctantly to earth.

Pigeons are wary birds with 360-degree vision; the mist clothed the sparrowhawk, but betrayed them.

Soil→grain→pigeon→sparrowhawk. One way or another, we're all living off the earth.

FOUR SEASONS IN one day. Rain, wind, sunshine, snow; the snow sufficiently heavy to blow up against the Chemical Brothers' hedge in scallops, and for the partridge to take shelter in the lee of the lane hedge. They stand on one leg, puffed-up balls, skittles all in a row.

In the snow by the ditch is an obscene ash smudge. I poke the skin and spikes of the hedgehog (there is nothing more, nothing less remaining) with my wellingtoned foot. The extraction of every edible part of the hedgehog is a wonder of butchery.

With the temperature consistently below 10 degrees centigrade, most of the Tiggywinkles have retreated to their hibernacula.

This one was out of time.

Hedgehogs have a reputation for cleverness. Or they do among poets. In the ninth century the Chinese poet Chu Chen Pu wrote:

> *He ambles along like a walking pincushion,*
> *Stops and curls up like a chestnut burr.*
> *He's not worried because he's so little.*
> *Nobody is going to slap him around.*

The Greek lyric poet Archilochus concurred on the inviolability of the hedgehog, writing: 'The fox knows many things, but the hedgehog knows one big thing.'

The six thousand or so spines on a hedgehog are

protection and, oddly, shock-absorbers. While a hedge-hog can climb up a tree, it cannot climb down one; to descend, it rolls into a ball and falls to the ground unhurt, cushioned by its angled spines.

It is a good trick. Yet the red fox has a better one. Around the remains of the hedgehog are flurries of fox paw marks; what the fox has done is roll the curled hedgehog into the ditch; to save itself from drowning, the hedgehog has forfeited its armoured protection and presented its head to the fox. Fatefully.

But the snow is traitor to the fox, friend to me. Reynard's footprints in the mauve snow of evening are a trail of evidence. The fox has been breaking and entering Flinders by climbing up and over the wooden field gate.

Just before dark I return and string barbed wire along the gate top.

Checkmate.

23 NOVEMBER At home, on the yard, fiddling with the fuel pipe of the International tractor by torchlight. Down in the valley, somewhere by the brook, a wail tears the night apart.

A rabbit has been taken by the fox.

Something about the incident, which is hardly unusual, unnerves me. I don't know if it is a premonition, or a message passed along the tendrils of existence which connect all things.

What I do know: I go to Flinders the next morning

with a heavy, prepared heart. Sure enough, one of the hares has been caught by a predator, a fox in all likelihood. Foxes are so-so as sprinters; as arrangers of ambuscades they are top drawer.

Wisps of brown hair are caught on stubble spikes. Looking closer I can see white in the hair. This winter, the old jack hare, he with the torn ears, was entering the winter of his life. Hares in the wild rarely live beyond three.

Hares go white with age, and their lip-split (hare lip) becomes more pronounced.

That night: sleet on Flinders. Watching mice and voles scampering for seeds in the stubble. The white chest of a barn owl – long time, no see – appears above my head. I could touch her. She starts her hunting; her soft wings caress the ground into silence and stupor. But the orange light of a tractor on the lane flashes across the field and beams her away.

No killer wants illumination.

3 DECEMBER Flinders. Starting my day here early, at 6.30am. There is a Venn diagram overlap of audio, when both night birds and day birds sound. The night birds are two owls, the first day birds two crows.

6 DECEMBER The winter sun is a flat blade of white on the horizon. Blackbirds are spinking, a charmless sound, thoughtless, done by rote.

You can take the boy out of the country, but you can't take the country out of the boy. My music taste is the despair of my family: I do the familiar near shore of British classical music (Butterworth, Vaughan Williams, Parry, Purcell); the French crooners Trenet, Aznavour, Bécaud (my stepmother taught French); Coldplay (nice boys' pop); The Clash (obvs) . . . And I have a soft spot for The Wurzels.

Penny, archly, on the subject of The Wurzels. 'Do they have any dates in Dalston? Brick Lane?'

She has a point: The Wurzels' tour is Stroud, Malmesbury, Barnstaple, Westward Ho!, Dartmouth, Bere Regis, Exmouth, Worcester, Cirencester, Bristol.

I'd be OK in Cirencester in the Cotswolds, I'd be perfectly camouflaged, since my habitual off-duty wear is a blue gilet, aka a 'Cirencester life jacket'.

Cirencester is sold out. So it's The Tunnels, Bristol, which is odd because I've not been to a club in Bristol since I was a postgraduate student.

It turns out I'm OK in Bristol Tunnels too because the audience is Zummerset, up for the night, up for a good time. 'I Am A Zider Drinker', 'I've Got A Brand New Combine Harvester'.

I've never had a better worse time. The people around me at the bar are real country; check shirts, accents that would saw wood. There are kids at the front wholly unable to dance, entirely able to drunkenly zing along; the Young Farmers' Club on an excursion. (Been there, done that.)

Of course, The Wurzels have scribed the greatest innuendo in pop: 'I drove my tractor through your haystack last night'.

Oo-ar-oo-ar-oo-ar.

We're all having fun. And yet some of the joy is the manic relief of the excluded, the forgotten, finding a voice.

Among the disappearing species in the countryside are country people. All the local casual work on the land is done by eastern Europeans who live apart, in caravan parks on farms; the Spar at Winter's Cross stocks no British beers or lagers, only Polish ones; there is no housing for local farm workers, so they have to commute to the countryside from Ross and Hereford; almost everyone who moves to the country to live is urban middle-middle class, and in the morning they travel in the other direction.

Wurzels albums include *Never Mind the Bullocks* and *A Load More Bullocks*.

*

WINTER, MID-1980S, university holiday: I'm on my Kawasaki 200 (an upgrade from the Honda 125) and on my way to Tenbury Wells to visit my maternal grandparents.

I pass the Talbot pub, and there on the front car park are the clan. Poppop is falcon elegant: Viyella check shirt, twill trousers, green knitted tie, fawn rain-coat flapped back by one casual hand in a trouser pocket. Sort of facing him, shiny-eyed and blotchy-faced with beer and whisky-chasers, his sons-in-law, his grand-sons, all in identikit check 'n' tweed, their brogued feet apart, two hands in front trouser pockets, but – and this is absolutely crucial – thumbs out.

It was the Temeside Agricultural Society Christmas lunch or some such.

I did not stop. We have been many things in my family – farmers, gamekeepers, courtiers, farm labour-ers, parsons (a lot of those), MPs, soldiers, teachers, reeves – but nobody before me had ever done anything as utterly non-vocational as History.

I was going one way, my family another.

Or so it seemed at the time.

Within a handful of years, I was back home, so the memory, actually, is not about leaving the land. (As I always say, 'You can take the boy out of Herefordshire . . . but not for very long.') The memory – and it has taken me years to realize this – is about the land leaving us. As with the hop-yard recollection, it is the background detail, barely

282

observed at the time, that dates the picture and consigns it to the unreachable past.

On the car park of the Talbot, the back door of a Land Rover is open, and there is a small pile of fur and feathers. One of my cousins, I learn later, spent a couple of hours the day before rough shooting on a neighbour's farm by Kyre Park. Just him and his cocker. In two hours he had bagged a couple of hares, six rabbits, two brace of grey partridge, a few 'woodies' (wood pigeons), two cock pheasant, a snipe, a jack snipe, two woodcock and a plover. He'd brought the haul to the pub to distribute among the family as Christmas presents.

None of the animals or birds had been reared for shooting. They just lived wild in the fields and spinneys.

I don't think I know of a farm in east Herefordshire or west Worcestershire that would now produce such a bag, in variety or in size, in two hours of rough shooting. Or, if the farm did hold such bounty, I doubt whether a 'gun' would shoot with an easy conscience and not be nagged by the worry that the killing of the hare or partridge might produce a local, ecological tipping point.

13 December Flinders. Relentless drizzle, which fails to stop long-tailed tits picking through a wild rose bush, from which the leaves have fallen, leaving a wire framework hanging with red jewels, as shop-display tasteful as anything in Bond Street.

283

Until the calendar reforms of Pope Gregory XIII in 1582, the winter solstice used to be on 13 December, St Lucy's Day. St Lucy was the girl for the job. Her name is derived from the Latin *lux*, meaning light.

Across the bare, wet earth of Flinders, worms lie side by side, shameless, conjoining, mating. In the wood a vixen shrieks her mating call. There is the promise of birth in nature as well as in Christianity in the midwinter month of December.

14 DECEMBER Flinders. Out at 11pm in frost. There's a rustle in the hedgerow; the dry floor acts as a drum skin, so the rat-sized noise by my feet is just a tiny snouty shrew. Poor shrews, ever active. There is no hibernation for them, only endless eating to keep alive, although they do have a clever winter-survival trick: they actually decrease their overall body size so there is less to maintain.

A tinnitus fox woo-woos repetitively, far off across sleeping land, somewhere over by Garway Hill.

I've put thirty Hebridean sheep in the wood. These are Viking-era ovines. Small, horned, black, they know what to do in woodland, because they browse as much as they graze. The Hebrideans are here to eat brambles, a surprisingly evergreen plant for a supposedly deciduous one. Brambles are wonderful for nature, up to a point. At the top of the wood the brambles have gone hegemonic. They are a domestic invasive species, creeping over the grass ride, crawling over the woodland flowers. They are taking over.

Sheep in woodland? It is a modern self-fulfilling fallacy that livestock, with the exception of pigs, have no place among trees. I run cattle through the woods, sometimes put the horses in too (with the exception of Zeb, my horse; he's from the Argentinian pampas and won't go anywhere which isn't flat and open). Livestock knock bits off, turn things over, bring life. And if they are old native breeds they will find much to eat, and the change of diet brings health. Either by instinct or by knowledge passed down in the DNA, animals possess an ability to select plant remedies for themselves. Welsh farms always used to have a *cae ysbyty* or 'hospital field' into which animals would be turned out when sick in the expectation they would find the remedy to cure themselves. It can work. A couple of years back, our miniature Shetland was playing with the Connemara and got a nasty kick to the shin. The Shetland went straight to the willow tree, and over the next twenty-four hours stripped off the bark and leaves and ate the lot. Willow contains salicin, the raw material of acetylsalicylic acid (aspirin), pain-killer and anti-inflammatory. Then again, the pony may have been eating willow because he is called Willow.

Moonlight settles on the floor of the wood as I shove open the gate. The de-foliated trees are silhouettes. This is nature in winter binary bleak. Black and white.

On along the shining path, between the old

285

pollarded alders. Under the hard glare of a December moon you can see things for what they really are. The alders are giant dark hands, thrust up through the earth, clawing for air.

The sound of a stick snapping underfoot is enough to scare the pigeons out of the oaks.

Deep into the wood now, past the shining pillars of of beech, the ground dry here. Weaving in and out of the trees, into shadow then out into the moon's glare. Black and white. Ahead, the bob-flash of a running rabbit.

The sheep are standing, alert; petrified ornaments. I call to them, 'Sheep!' They recognize my voice, animate, and the tamer ones edge forward. A few of the ewes are on heat: the ram is flehming – pulling back his upper lip to flash his gleaming teeth. To you and me this looks like gurning; to a female sheep it is George Clooney bedroom-smiling.

Leaning against the cherry tree I, a shepherd, count my flock by night. Cherry bark has thin rough rings; with my index fingernail I notch a ring per sheep. The tree is my abacus.

Nature-watching beside a flock of sheep is always the smart move; ovine odour camouflages human scent. So I am not entirely surprised on this night of wonders to see the badger shambling along the path towards me.

Like the shrew, the badger does not hibernate.

Closer he comes. He snuffles. He squats. He scents. All this is done with a casually proprietary air. There is menace too. Those bad-boy black stripes on the head are, in the jargon, aposematism, or warning colouration. Brock has teeth.

By now the badger is within touching distance. But he never sees me, and he shuffles on past, nose to ground, to vanish into the black-and-white night.

Walking back across Flinders, the moonlight makes shadows of three animals with long ears, longer legs.

I've seen hares by moonglow, and I've gazed into the heavens.

I've felt the true peace of the earth.

18 DECEMBER Flinders. A snow flurry overnight has coated ploughland, obliterating every trace of earth.

The two leverets race around in the centre of the field, scuds of snow flying up behind them. They pass so close to me that I hear the thud of their pads.

From the tracks in the snow I can see that the hares have now colonized the top paddock; so they now have the run of three fields. The Empire of the Hare goes on.

BOXING DAY NIGHT. A maudlin mood beside the log burner. I look at the pictures in my 1970 *Reader's Digest Book of British Birds*. Run my fingers over the plates in longing. Because that is all I can do. I never see half of the birds in the book in the wild any more. Turtle doves?

Quail? Cirl buntings? Corn buntings? I cannot remeber when I last saw any of these.

But then I think. One field, just one field, made a difference.

If we had a thousand fields . . .

The End of the Affair

THAT WAS NOT THE end at Flinders. Not quite.

In accordance with the terms of my tenancy, I laid the field down to grass in the following March. Extraordinarily, birds were still finding seeds among the 'weeds' of the wildflower borders in this month, meaning the field had carried them through an entire winter.

The ley, together with the lack of turning of the soil, effectively killed off all the arable flowers in a season, except in the gateways where pedestrian and farm traffic continued to churn the soil. Here poppies and corn chamomile, scarlet pimpernel and speedwell continued to thrive.

But I'd long had my eye on the derelict cottage garden which came with the tenancy. The garden was a monoculture of brambles growing over carpet someone had laid down as a weed-suppressant so long ago that the fabric had adhered to the earth.

Brambles abounded in the wood next door; I made the calculation that nature would benefit if I utilized the half-acre garden as a small cereal field, a Flinders in miniature.

Also, I wanted to know how little land is needed to make a qualitative difference in farmland conservation. How low in acreage can one go and be beneficial?

For long January days I prised up the carpet with crowbar and spade, and cleared the brambles with a brush-cutter. The glistening soil underneath was rich, black and fertile; and crawling with worms and centipedes. Then for six weeks I 'folded' (fenced) sheep on to the plot, with straw bedding.

There was a wait of ten weeks until the ground dried sufficiently for me to rotavate the garden, it being too small to plough with a tractor. If the sheep had fed the soil, they had compressed it too, and walking behind the Camon 2000 rotavator while its Boudicca blades bit into the soil was akin to holding down a bucking mechanical pony on the bucking deck of a ship.

'Delirium' has its roots in the Greek for the madness that comes from not ploughing a straight furrow; Ulysses was 'delirious' when he took his plough to the beach to try and cultivate it.

I developed 'rota-delirium', the brain-shaking insanity that comes from rotavating half an acre.

Willow and I harrowed the land, three times. Once for tilth, twice to remove unwanted wild plants, notably horseradish and docks, whose roots extended for miles under the carpet.

On half the plot I planted white millet, as cover and food for birds. On the other half I drilled wheat – this time in straight rows, which could be hoed. I cast some poppy, corn chamomile, cornflower and corn marigold seeds throughout the plot; most seeds, however, went in one-metre-wide wildflower borders around the plot, with a generous two-metre border separating the two halves.

The wheat I planted was special: a 'heritage' spring wheat called April Beard, obtained from the Brockwell Bake charity in London, which encourages people to turn lawns into wheatfields.

When mature, April Beard is long (120cm) and hollow tubed. Once upon a time children drank their milk through natural 'straws' made from wheats such as April Beard.

The cottage garden was uphill from Flinders and post-Flinders, and in a sense gave me a plan view on, and a retrospective of, the original experiment.

Once again I erected a bird table, although many birds were already habituated into visiting the Flinders area.

The aphorism is 'Birds of a feather stick together'. More accurate is 'Birds magnetically attract birds'. Birds flying overhead would descend to see what other birds were eating. From the garden I could see that birds were flying in from more than a mile away.

Although Flinders itself had gone to grass it was still chemical-free. The moles moved into the field the same March I rotavated the garden.

A male pheasant quickly laid claim to the garden; he was a venerable warrior. No bird has more guises. This one in his burnished bronze armour was Achilles.

In April a pair of red-legged partridge nested in the emerging wheat; since the gate to the paddocks down to Flinders had a mesh bottom I inverted it so hares could clamber in between the horizontal bars; sheep could not.

By June, the wheat and the millet were a foot tall, and the wildflowers were blooming. The red-legs had seven chicks. Collared doves, wrens, blue tits, blackbirds all nested in the garden.

In the evenings wildlife seeped in for safety and for food. Four more hares fed on the young wheat. Whole families of birds feasted on the emerald shield bugs which decorated the millet; it was a good time for the pied wagtails. The kestrels added the garden to their predatory round.

I cut the wheat late, on 19 September, when the air was heavy with that end-of-the-affair repose. It was the day the last swallow left. (Swallows always announce their departure by noisy massing; I have never seen swifts leave, they simply slip away when one's back is turned.) Autumn's flames were already licking the pear tree.

*

I have scythed hay by hand, yet I venture to say that one fails to understand the vigour of the countryman before the coming of the machine age until one has reaped corn with a sickle.

Richard Jefferies witnessed reaping by hand in its Victorian prime:

No one could stand the harvest-field as a reaper except he had been born and cradled in a cottage, and passed his childhood bareheaded in July heats and January snows. I was always fond of being out of doors, yet I used to wonder how these men and women could stand it, for the summer day is long, and they were there hours before I was up. The edge of the reap-hook had to be driven by force through the stout stalks like a sword, blow after blow, minute after minute, hour after hour; the back stooping, and the broad sun throwing his fiery rays from a full disc on the head and neck. I think some of them used to put handkerchiefs doubled up in their hats as pads, as in the East they wind the long roll of the turban about the head, and perhaps they would have done better if they had adopted the custom of the South and wound a long scarf about the middle of the body, for they were very liable to be struck down with such internal complaints as come from great heat. Their necks grew black, much like black oak in old houses. Their open chests were always bare, and flat, and stark, and never rising with rounded bust-like muscle as the Greek statues of athletes.

The breast-bone was burned black, and their arms, tough as ash, seemed cased in leather. They grew visibly thinner in the harvest-field, and shrunk together – all flesh disappearing, and nothing but sinew and muscle remaining. Never was such work.

Silica in the wheat stalks constantly blunts the sickle blade.

After reaping came the hand-tying of sheaves, an art in itself, as Jefferies explained:

The gleaners had a way of binding up the collected wheatstalks together so that a very large quantity was held tightly in a very small compass. The gleaner's sheaf looked like the knot of a girl's hair woven in and bound. It was a tradition of the wheat field handed down from generation to generation, a thing you could not possibly do unless you had been shown the secret – like the knots the sailors tie, a kind of hand art. The wheatstalk being thick at one end makes the sheaf heavier and more solid there, and so in any manner of fastening it or stacking it, it takes a rounded shape like a nine-pin; the round ricks are built thick in the middle and lessen gradually toward the top and toward the ground.

As with many a job down on the farm, appearances are deceptive. Straw scythed by hand is often irregular in length, and lies scattered, uncomely, on the ground, and thus needs to be fought into shape. If the resultant sheaf is loose, it falls apart; if it is tight, it will not dry. The straw scratches, the nettles sting, the thistles prickle. Hand-binding is one of those impossible rural jobs which require gloves, though the self-same gloves prevent the very nimbleness needed.

Only after I had tied twenty or more sheaves were mine in any way pretty, except for the wildflowers accidentally tied into their hair. The quarter of an acre of stubble, plus the millet, managed to hold the red-legs, who were joined by their extended family of twenty or more in a winter covey.

Red-legged partridges, unlike grey partridges, will perch off the ground. Once, in December, when I went to check on the garden I beheld every birdwatcher's and every country person's dream: a partridge in a pear tree.

I drove past Flinders recently, and looked over the gate. The day was darkling, so I turned on the Land Rover's headlights.

There were still hares there. Running. Dancing.

A Ploughland Reading List

Eve Balfour, *The Living Soil*, 1943

B.B. (Denys Watkins-Pitchford), *Letters from Compton Deverell*, 1950

Adrian Bell, *Corduroy*, 1930

John Clare, *The Shepherd's Calendar: with Village Stories and Other Poems*, 1827

William Cobbett, *Cottage Economy*, 1821; *Rural Rides*, 1830

John Stewart Collis, *The Worm Forgives the Plough*, 1973

William Cowper, *Letters*, 1836; *Selected Poems*, 1984

Roger Deakin, *Notes from Walnut Tree Farm*, 2008

Department for Environment, Food and Rural Affairs, *Wild Bird Populations in the UK, 1970 to 2014*, 2015

Henry Ellacombe, *Plant-Lore & Garden-Craft of Shakespeare*, 1884

George Ewart Evans, *The Horse in the Furrow*, 1960; *The Leaping Hare*, 1972

'Romany' (George Bramwell Evens), *A Romany in the Fields*, 1929

Stella Gibbons, *Cold Comfort Farm*, 1932

Nick Groom, *The Seasons*, 2013

H. Rider Haggard, *A Farmer's Year*, 1899

Otto Herman and J. A. Owen, *Birds Useful and Harmful*, 1909

W. P. Hodgkinson, *The Eloquent Silence*, 1947

W. G. Hoskins, *The Making of the English Landscape*, 1955

Richard Jefferies, *The Life of the Fields*, 1884; *Field and Hedgerow*, 1889 *Wild Life in a Southern County*, 1879; *The Story of My Heart*, 1883

William Langland, *Piers the Ploughman*, 1959

Robert Macfarlane, *Landmarks*, 2015

John McNeillie, *Wigtown Ploughman*, 1939

Peter Marren, *Rainbow Dust*, 2015

E. M. Nicholson, *Birds and Men*, 1990

Oliver Rackham, *The History of the Countryside*, 2000

Henry Stephens, *The Book of the Farm*, 1844

A. G. Street, *Farmer's Glory*, 1932

Edward Thomas, *The Last Sheaf*, 1928; *The South Country*, 1909; *Collected Poems*, 1953

Thomas Tusser, *Five Hundred Points of Good Husbandry*, 1573

Brian Vesey-Fitzgerald, *British Game*, 1953

Henry Williamson, *The Story of a Norfolk Farm*, 1941

P. J. Wilson and M. King, *Arable Plants: A Field Guide*, 2003

Esther Woolfson, *Corvus*, 2009

A Ploughland Music List

Blue Oyster Cult, '(Don't Fear) The Reaper', 1976

George Butterworth (words by A. E. Housman), 'Is My Team Ploughing?', 1911

Incredible String Band, 'Douglas Traherne Harding', 1968

Small Faces, 'Song Of A Baker', 1968

The Wurzels, 'The Combine Harvester', 1976

Jane Montgomery Campbell, 'We Plough the Fields and Scatter', 1861; original German by Matthias Claudius, 1782; music by A. P. Schulz

Ivor Gurney (words by Seosamh MacCathmhaoil), 'I Will Go With My Father A-Ploughing', 1921

Pink Floyd, 'The Scarecrow', 1967

Jethro Tull, 'Heavy Horses', 1978

Henry Purcell, 'When I am Laid in Earth' (Dido's Lament), 1689

Traditional, 'The Ox Plough Song'; 'The Painful Plough'; 'Speed The Plough'; 'We Are Jolly Good Fellows That Follow The Plough'

Traffic, 'John Barleycorn Must Die', 1970

Robert Wyatt, 'Pigs . . . (In There)', 1999

Peter Pears, 'The Ploughboy' (traditional)

Ralph Vaughan Williams, *The Lark Ascending*, 1920

Edward Elgar, 'Caractacus', 1898; Cello Concerto, 1919

Acknowledgements

As ever: Susanna Wadeson, Julian Alexander, Sophie Christopher, Ben Clark, Lizzy Goudsmit, Kate Samano, Phil Lord, Geraldine Ellison, Patsy Irwin, Penny Lewis-Stempel, and Edith Swan-neck. Plus Deborah Adams, Josh Benn, Rob Pryor, Phil Miller, Margaret Yarnold, Kathryn and Nicholas Fox, Sue and Rich Francis, Felix and Richard Jones.

And to my parents, grandparents – thanks for the (country) memories.

I have changed some names to protect the guilty.

John Lewis-Stempel is a writer and farmer. His many previous books include *The Wild Life: A Year of Living on Wild Food*, *England: The Autobiography*, *Six Weeks: The Short and Gallant Life of the British Officer in the First World War* and *Meadowland*, which won the Thwaites Wainwright Prize in 2015. John writes for *Country Life* and won the BMSE Columnist of the Year Award in 2016. He lives on the borders of England and Wales with his wife and two children.

The Wild Life
A Year of Living on Wild Food
John Lewis-Stempel

For a full year John Lewis-Stempel lived solely on food he had hunted, shot, caught or foraged from the fields and copses of his small seventeenth-century Herefordshire farm. The experience brought him closer to nature, closer to his history and family and most importantly it brought him to a better understanding of himself.

'A great book: tough and funny, metaphysical
and earthy, passionate and honest'
ROBERT MACFARLANE

'Often funny, always passionate, this is a fascinating read'
BBC COUNTRY FILE

'Beautifully written. The closest thing you will find
to poetry in prose'
PAUL BLAZARD, HAY LITERARY FESTIVAL

'A fascinating account of each month as he tracks, kills and gathers
what he needs to stay alive, and a meditation on survival and our
connection to the land . . . timely and compelling'
JASON WEBSTER, SUNDAY TELEGRAPH

Meadowland
The Private Life of an English Field
John Lewis-Stempel

'To stand alone in a field in England and listen to the morning chorus of the birds is to remember why life is precious.'

In exquisite prose John Lewis-Stempel records the passing seasons in an ancient meadow on his farm. His unique and intimate account of the birth, life and death of the flora and fauna – from the pair of ravens who have lived there longer than he has to the minutiae underfoot – is threaded throughout with the history of the field and recalls the literature of other observers of our natural history in a remarkable piece of writing that follows the tradition of Jeffries, Mabey and Deakin.

'A rich, interesting book, generously studded
with raisins of curious information'
THE TIMES

'A magnificent love letter to the natural world, full of wisdom and experience, written with wit, poetry and love. This is, in fact, one of the best five books I have ever had the privilege to read'
TIM SMIT, THE EDEN PROJECT

'Fascinating . . . books have been written about entire countries that contain a less interesting cast of characters'
TOM COX, OBSERVER

'Engaging, closely-observed and beautiful'
BEL MOONEY, DAILY MAIL